SMELLS, ODORS, AND SCENTS FROM HEAD TO TOE: IN PERSONAL RECOVERY AND ECSTASY!

By Ronald Alan Duskis, D.C., B.A. Zoology, and B.A. Theology

toExcel

San Jose New York Lincoln Shanghai

Smells, Odors, and Scents from Head to Toe
In Personal Recovery and Ecstasy!

Published by toExcel
an imprint of iUniverse.com, Inc.

For information address:
iUniverse.com, Inc.
620 North 48th Street
Suite 201
Lincoln, NE 68504-3467
www.iuniverse.com

ISBN: 0-595-00530-6

Printed in the United States of America

TO ALL OF HUMANITY: IN THE HOPE OF A DELIGHTFULLY SMELLING WORLD OF PEOPLE!

AND TO MY FAMILY FOR THE LOVE, RECOVERY, AND ECSTASY WE ALL SHARE!: MY WIFE AND MY TWO DAUGHTERS: CHRISSY AND CHARISSA!

ACKNOWLEDGEMENTS

Many people have contributed to the success of this writing, and I desire to express deep thanks for their help.

First, I am especially grateful to my Dad and Mom for all the support they have given me throughout the years.

Thanks are also due to all the people who have given their testimonials and/or comments on their personal experiences with pleasant and/or unpleasant odors. I also want to thank all of you who gave me articles or other information on odors. And I want to thank all of you who will want to participate in upcoming writings of this book and other books I am writing.

Thanks are also due to any Doctors, companies, and/or Health Food stores that were willing to discuss their services and offer them to the reader in hope of the reader delighting in odor recovery and/or ecstasy! Thanks are also given in advance for any other professionals or companies that would like to participate in the next planned writing on odors.

And special thanks are also given to all of you, the readers! Thank you for taking the time to improve your life on your journey to wholeness!

I would also like to thank the following people: John Miller, for helping me to set this writing on the computer: he spent time teaching me to skillfully use the computer at his home and at the Computer Lab at the Ambassador College in Big Sandy, Texas. Also, a special thank you to two Ambassador College students for all their help in the Computer Lab helping me with this writing whenever I needed them: Paul Ramcharit and Pedro Reynoso. Sheryl Boyle, an Ambassador College student helped me to locate some reference books on odors-Thank you Sheryl! Also, my daughter, Charissa Duskis, thank you for your help in the library also!

TABLE OF CONTENTS

ACKNOWLEDGEMENTS...PAGE 1.

INTRODUCTION..PAGE 5.

TESTIMONIAL SHEET FOR NEW TESTIMONIES......................PAGE 6.

EXPLAINATION OF TESTIMONIAL SHEET..........................PAGE 7.

HOW THIS BOOK CAME ABOUT AND ITS NEED FOR TODAY...........PAGE 9.

HEALTH IN THE MOUTH: MOUTH RECOVERY.......................PAGE 29.

ECSTASY OF MOUTH ODORS.....................................PAGE 51.

HEALTH IN ARMPITS: ARMPIT RECOVERY........................PAGE 64.

ECSTASY OF UNDERARM ODORS..................................PAGE 75.

HEALTH IN THE GENITALS: GENITAL ODOR RECOVERY............PAGE 83.

HEALTH IN THE GENITALS: GENITAL ODOR ECSTASY.............PAGE 89.

HEALTH IN THE FEET: FEET RECOVERY FROM ODORS.............PAGE 94.

HEALTH IN THE FEET: FEET ECSTASY..........................PAGE 97.

HEALTH FOR THE SKIN: SKIN RECOVERY FROM ODORS...........PAGE 100.

HEALTH IN THE SKIN: SKIN ODOR ECSTASY....................PAGE 106.

BIBLIOGRAPHY...PAGE 114.

INTRODUCTION

Odors, enjoyable and not so enjoyable, have entered our noses from various sources. The joy and excitement of wonderful odors has been and continues to be the experience of generally all people everywhere. Fortunately, even the unpleasant odors can be helped or eliminated by overcoming the causes that make them.

Whether you want to enhance the pleasant odors you possess or want to overcome some odor(s) that have bothered you or your neighbors, this book is for you. I also am constantly learning new facts from books, people's life experiences, and professionals. I also desire dearly for <u>any</u> reader to send me any information they may now possess that is new to the information in this book or that they can add to what is already in this book for future editions by writing to me immediately with their signature in the place provided on the next page to give me permission to let me use what they write in my book. That page may be copied and given to friends who also wish to participate in this book. I have other books that I am collecting testimonials on also. This process will hopefully continue for years to come; so please send in your testimonials <u>now</u>. I would like to thank everyone who has helped make this book a reality, and I want to put your name in the book if you would like it to be there (just indicate that or other choices on the testimonial page on the next page of this book). Also, if you happen to be reading any articles, books, magazines, newspapers, etc. and come across any information that I could use in future editions of this book, please send these to me copied if it is 5 pages or less or else tell me where I can find it. Many thanks to all of you for your taking part in this writing of a subject that is so important in our daily lives!

To anyone desiring to participate in this project given below:

I understand that Ronald Alan Duskis is gathering real life experiences and knowledge dealing with any odors found, pleasant or unpleasant, from head to toe or in your environment. The following are my experiences or knowledge dealing with this subject that may be used in any of his literature or publications in full or part anytime with my complete consent and blessing in his using it without any money ever to be given to me or anyone else.

Signed_____Date_____

Print Your Name_____

Address_____

Phone Number_____

_____
_____
_____
_____
_____
_____
_____
_____
_____
_____
_____
_____
_____
_____
_____
_____
_____
_____
_____
_____
_____
_____
_____
_____

_____
Signed_____Date_____
Place your <u>signature</u> in the following desired:
1. I want my name and address in your printing_____.
2. I only want my name printed in your work_____.
3. I only want my initials and city and state_____.
4. I definitely want my phone number listed_____.
5. Special further instructions as needed_____

_____.

To all people participating in my writing on odors:

Thank you for participating in one of the most exciting and needful writings in the decade of the 90s. Your help will be greatly appreciated by the many people who suffer minute by minute with odors that are unpleasant. Your help will possibly reach not only thousands but hopefully millions in the months and years to come. So please take your time in filling out the page that has space for your experiences or knowledge. Copy that page if you need more room. And please do not forget to fill in all signature spaces with your signature, and answer any questions that pertain to you on the bottom of that page from numbers 1 through 5. Please return the page to me as soon as possible so that your writing may be included in my writing. Once again, thank you for your help and participation!

Remember, that by participating and helping, you can help your fellow man not only smell pleasant, but you may save yourself the uncomfortable situation of smelling someone who you did not help! Also, by helping others, if enough people just read this book, you and others will experience a world where we can wake up curiously wondering what delightful types and combinations of odors we will experience today and days to come. Also, the fear of not knowing what odor may come out of someone's mouth or arm pit or other body part will be gone!

And who knows just how many new friends and relationships can be attracted to you through pleasant odors you have always wanted to give to others but did not know a book to find the answers in. How curious you will be to see the varied comfortable reactions to you when you learn from this book how to artfully change your pleasant odors daily as you desire.

Once again, you will no longer have to fear not knowing the reaction of people to your bad odors since they should be eliminated hopefully forever with help from this book and your Doctor as needed. Please note that all suggestions given in this book should be checked with your Medical Doctor in case of allergies or other health related problems.

So give to your fellow man your best experiences or knowledge dealing with any odors, pleasant or unpleasant. These can deal with odors of the head, mouth, skin, arm pits, private parts, feet, and environmental. For the sake of those suffering from foul smelling stools and urine, give some experiences and knowledge also if you know of any.

Thank you again for your participation and support,

May 15, 1991

Ronald Alan Duskis   Date signed

ABOUT AND ITS NEED FOR TODAY

This section can be skipped unless the reader is curious as to the importance of odors, pleasant and unpleasant, personally and culturally. A brief look at the anatomy and function of the sense of smell will also be presented in this portion of my book. I hope that you all enjoy it as much as I do. What an exciting subject! It can bring the creativity and curiosity in all of us to a higher level of participation in our lives; which in turn, can help us to better work within our environments for the pleasure of all people's lives as we interact one with another. Come with me now and see what I mean!

I have had an extreme interest in pleasant and unpleasant odors personally due to my history. Since earliest childhood I had wondered why some things smelled so marvelous and others so awful! I learned early in childhood that I was not to speak too much about the awful smells, that it was like a forbidden and banned subject. It was like there was a secretive aspect to unpleasant odors that must not be mentioned if someone I knew owned that awful smell. As a child I found this humorous since it appeared that unpleasant odors came out of all of us humans, so why the shame and secrecy about it? I was puzzled by the attitudes of people towards certain foods like garlic in which some would like the smell and others would not! Some people would like the smell of a person's perfume, and others like myself did not. I grew up wanting to know why people react to odors in the way they do; some shame others while others helped people overcome their undesired smells; and still others even helped people to smell wonderful! My interest intensified when I found myself suffering from bad breath at an early age of about 7.

My father, I was told, wanted to be a Medical Doctor. I learned that of the three sons he had, I was number 2, he chose me to be a Doctor, if I showed the interest only, He guided me in this direction due to my extreme desire towards solving the mysteries of health and dis-ease, as an interdisciplinary study with other fields of study. Inter means "between" and discipline means to self-control within a study once the principles of that study are known for self-good as well as the good of everyone. More will be said of this in the coming chapters of this book, and how it can help you the reader.

Due to the bad breath, that I remember came on me rather suddenly about the age of 7, I did not date much nor talk to people if they got too close to me. Fortunately, this did work together for good though since it stirred me to study and emphasize the sciences early in life, a study that *has never ceased to excite my curiosity. I excelled in my school work,*

working diligently to find answers to life's questions about health and

illness; and its effects on our interpersonal relationships with one another

and our culture. This has made my journey in life interesting to say the

least. I have gone to college for years and have received degrees in

Zoology, Chiropractic, and Theology; and I minored in Chemistry. I was

Pre-Med at UCLA from about 1968 to 1969, and I did some help in

research there, at that time, on cell multiplication changes in Euglenoid

cells due to our changing the chemical environment.

After graduating from High School in 1965 where I had excelled in

Chemistry and other sciences, I went to Jr. College in which I emphasized

chemistry and zoology (the study of all animals). I graduated there in

1967 with an A.A. Degree in Chemistry, if my memory serves me correctly.

I was highly inquisitive about the reason for the odors in animals and

humans at that time in relationship to the chemical structure of the odor.

I read that certain structures and functional groups would give certain

odors. Robert Burton's book, The Language of Smell, on pages 23-26

shows this appears to be the case: "Molecular structure was postulated as

the basis of smell two thousand years ago. In 27 B.C. Lucretius wrote,

'such substances as agreeably titillate the senses are composed of smooth,

round atoms,'...and further work has shown more close correlations between other groups of similar odours and the size and shape of their molecules. Amoore concluded that these groups are the primary odours and that there are seven of them. In the simplest form of the theory a molecule of each primary odour is seen as fitting into a 'site' in the membrane of the receptor cell as a key fits a lock, as follows: ...Primary odour...musky...shape of molecule...disc...shape of site...elliptical bowl but larger...putrid...shape of molecule...negatively charged...shape of site...positive charge...So, tissue in the nasal clefts contains seven kinds of receptors, each responding to the particular shape of the molecules of a primary odour...Primary odour (s)...camphoraceous...musky...floral..peppermint...ethereal..putrid...(and) pungent...To test Amoore's theory, a number of new chemical compounds were synthesized and predictions of their odour were made from stereochemical measurements of their shapes. The shapes and odours were found to tally and so confirm Amoore's theory."

Similarities in shape of the molecules for similar acting behaviors were created to be that way for a purpose. The purpose is for us to learn to build character. To learn to choose, while using self-control or temperance, the ways that will bring love, joy, peace, goodness, gentleness, etc. to all mankind's lives. These chemicals can be used to bring about ecstacy if used correctly; or can be used in such a way that will bring heartache or even illness or finally death if used in ways not intended. For example, Robert Burton shows the similarities of structure for the sex hormone testosterone with the following other chemicals, on page 111 of his book: "In 1948 L. LeMagnen, a French physiologist, discovered by accident that adult women were very sensitive to the smell of exaltolide, whereas man or young boys and girls either failed to smell it or could do so only feebly. A woman's sensitivity to exaltolide, and many other chemicals, varies through the course of her menstrual cycle. She is most sensitive at the time of ovulation and is relatively insensitive during the first months of pregnancy. Exaltolide is extracted from angelica, the root of which is used to flavor liqueurs such as Benedictine. It smells like, and is chemically similar to , civetone and muscone, the pheromones used by male civets and musk deer respectively. It is mainly the musky odours to which women are more sensitive than men. 'Boar-taint' of pork is another musk-like substance smelt more readily by women. It appears to be made from a musky substance secreted in the urine of boars and, as the name suggests, it imparts an unpleasant odour to pork made from uncastrated boars. The musky substance is a pheromone which causes a sow on heat to stand in the arched mating position when a boar approaches. A similar substance is found in human urine. A pattern can now be seen emerging. Women are sensitive to substances that have a musky odour, that are used as pheromones by various mammals and are often associated with the male animal. Moreover, these 'musks' are chemically related to the male hormone, testosterone." Thus odors can be used to attract the right or wrong person to us; therefore, I have found that I need to know what is a relationship worth building for right character sake through the use of odors, and which relationships I want to avoid as they only lead to eventual emptiness, physically, mentally, and spiritually. I found that the choice is mine!

At UCLA, from 1968-1969, I learned, amongst many other things, the importance of low concentrations of chemicals, and how they affect and relate to the human and animal bodies. Robert Burton gives this sensitivity in relation to odors in his book on page 108: "...just how sensitive is this much maligned sense. From anatomical and behavioral considerations it is clear that Man's sense of smell does not measure up to that of many animals, such as the domestic dog, but our noses are still extraordinarily sensitive. One olfactory receptor cell in the human nose requires only eight odorous molecules for stimulation and forty or more need to be stimulated for a conscious sensation of a smell. Sensitivity to such a small number of molecules explains how a man can detect the odour of a footprint on blotting paper and why one milligram of skatole (the odour of faeces) can render unpleasant the atmosphere in a hall 500 meters long, 100 meters wide and 50 meters high. The human nose can also distinguish a huge variety of odours. A skilled perfumer can distinguish up to 10,000 odours. He will have a range of 2,000 ingredients for use in his work and can readily identify half of these." And Steve Van Toller and George H. Dodd on pages 175-176 shows that these many odors affect our lives in many ways: "It would seem that there is strong evidence for a connection between odour and certain basic human drives such as food and sex, and responses such as awareness of danger. Many researchers have detected physiological changes in response to odour stimulation in areas such as blood pressure, muscle tension, skin temperature, skin conductance and brain wave patters...in general we do respond emotionally to odours...Certainly aromatherapy does work psychosomatically since it acts on both body and mind via the olfactory nerve tracts and the central nervous system."

I also studied the anatomy and physiology of humans and animals at UCLA between 1968 and 1969. Some important facts relating to odors in *this area would be important to consider by the curious and wondering and excited reader. Robert Burton covers some of this in his book. On page 22 he shows, "...that there are thousands of known odours...the sense of smell is extremely sensitive; even Man can detect contaminants in certain substances, such as menthol and saffrol, at much lower concentrations than can be detected by standard methods of the chemist. It seems that no more than eight molecules need to hit a receptor cell for a nerve impulse to be generated and impulses from forty receptors are needed for the smell to be perceived." On page 23 he says, "In the nose, it is presumed that odorous molecules impinge on the receptor membranes and distort the molecules in the membrane so that the sodium ions can pass through." Also, Robert Burton gives the definition of the science of smell: "Technically the science of smell is (osmics from the Greek osme=a smell) and odorous substances are osmyls. The act of smelling is olfaction from the Latin olfacere=to smell. The membrane in the nose is the olfactory membrane and the parts of the brain dealing with smell are the olfactory lobes." On page 9 he shows what a pheromone is: "An odour, or a smell or a scent, consists of molecules of a volatile substance that are carried in the air...The sense of smell is therefore, the detection of chemicals. In the language of smell, odours are secreted by* one animal and carried in air or water to the olfactory organs of another animal of the same species. the odours act as messengers telling an animal about its fellows...they are called pheromones (from the Greek pherein=carry and horman=excite) and they are analogous to hormones, the chemical messengers secreted by an internal gland and carried by the blood to affect distant parts of the body. The standard definition of a pheromone is that it is a substance which is secreted to the outside of the body and received by a second individual of the same species, in which it releases a specific reaction, for example, a definite behavior or a developmental process."

Robert Burton on pages 21-22 gives the anatomy of the tissue of smell or olfaction: "...there are four kinds of tissue cell. The sense cells, or receptors, form a layer of tissue called an epithelium in which each is surrounded by supporting cells that form a kind of packing, isolating the receptors from one another. Around the edges of the olfactory epithelium there are cells that secrete mucus and cells with cilia that continuously beat to propel the mucus over the olfactory epithelium...cilia greatly increase the surface area of the olfactory epithelium and are the site of interaction between odorous molecules and the receptors...odorous molecules brought into the nasal clefts dissolve in the mucus and somehow stimulate the receptors to create an electric charge, a process known as transduction." He also shows on page 20 that there are two cavities or clefts in the nose in which in normal breathing are hardly entered by the molecules of smell unless they are strong odors; but, if sniffing is done, and the olfactory epithelium which is yellowish is moistened with mucus, then weaker smells can be detected. There are 5,000,000 (five million) receptors in these two sheets for man. On page 3 he tells us that the membranes when put together are about the size of "two postage stamps".

And further at UCLA, I learned a remarkable fact about the olfactory system as here expressed in the book called *Perfumery The Psychology and Biology of Fragrance* (see Bibliography) on page 186: "Smell is the only sense which has direct contact with the brain...the limbic system...the 'smell brain'...serves as the central circuit for emotion, mood, motivation and sexual behavior. It can be stimulated directly through the sense of smell...It is thus understandable why the perception of certain fragrances...can be connected with a recollection of specific experiences, or why when we experience childhood odours, we simultaneously perceive past emotions." Thus on page 191 of this book is written: "...the perfume must correspond to the emotional perfume needs of a person...this is *virtually the basic prerequisite for the fascination of a perfume, and the selection of a perfume for continued use. If the radiance of the perfume does not meet one's emotional needs, it is at best considered interesting, but this interest will be short-lived." The book continues on page 192 to show that there are emotional perfumes and rational perfumes to fit ones mood needs; and that women will have many perfumes depending on the need of the period of time they are experiencing. For example, if they feel a need to be more introverted or tranquil, they can look for an emotional perfume.*

After UCLA, I taught at a private Junior High School and High School in the field of science and any other subject that was taught. This is because it was a school in which students could go at their own pace in all of the required subjects. I liked that method of teaching. I felt happy to see the students more comfortable at going at their own rate that pleased them since time was not a major factor. In that environment I learned to integrate the disciplines seeing the need for all kinds of studies and their applications in my life. I began to see the need to examine Chiropractic and Theology, two fields that I knew something about but not to the degree of discipline and integration as part of my being in my journey to wholeness. And all my life, since I was a little boy, I remember wanting, craving, to know how everything fits together in an interdependent way. I

kept wondering what life is __all__ *about, I did not want to leave any out that*

would bring a happy environment for all mankind.

The Chiropractic gave me some more missing parts to the puzzle. In

Chiropractic I learned that the nervous system, which is the governing

system, reacts to the way I live my life. In my first book, Back To Health,

I discuss the three major categories of principles I learned, while going to

Chiropractic College in Los Angeles, California, that affect the health of

my body. I called this the MAN principle: M is Mechanical. A is

Attitudes. And N is Nutritional. These are the three basic categories that

affect our physical health no matter what our beliefs are. For example, if

mechanically we get into a car accident we usually put out some of the

bones of the spine which affects adversely the governmental messages to

our organs and tissues including the olfactory cells for smelling pleasant

and unpleasant odors. If we have an attitude that tightens our muscles,

for example in the shoulder region because we are angry at someone,

then the tight muscles tug-of-war on both sides of the spine and pull the

bones of the spine out of position once again affecting detrimentally the

nerve supply to the area of smell, the mouth, armpits, groin, feet, skin,

and every other area throughout the body. Nutritional abuse does the same. From this I learned that I had better build character by building healthy mechanical, attitude, and nutritional boundaries.

During the years from 1970 through 1978 I taught at two different Chiropractic Colleges in California. I taught nearly every subject which further helped me to see the big picture in life, and most of the subjects were taught using medical textbooks. In Chiropractic I have heard that we nearly take every course that the Medical Doctors take except for subjects like Pharmacology. I heard that we take more neurology than they do, and that we take as many hours as they do in our going to college. I taught Philosophy, Psychology, Psychiatry, Anatomy, Histology (the study of tissues of the body), Physiology, Neurology, Health and Hygiene, Pathology, Clinic, X-rays, Physical Therapy, Geriatrics (study of the elderly), Pediatrics (study of the child), Obstetrics, Chemistry: Organic and Inorganic, Biochemistry, Bacteriology, Symptomatology, Dissection, Bacteriology Lab., etc.: all these subjects related to the study of smell to help me to write a more completely integrated, curiosity provoking, and exciting book. It should prove to be a book that can profit most everyone in knowledge, understanding, and wisdom on the subject. It is an important part of my journey to wholeness. It taught me to put together the sciences that I had studied in the past in High School, Junior

College, teaching Junior High and High School, and UCLA with these as a discipline that was interdisciplinary or integrated so that I could practice more self-control for the development of character.

In Theology, I graduated once in 1980 with a certificate, and I graduated with a B.A. in 1991, I learned how to be interdisciplinary with all studies. I went to a college that emphasizes the whole man and how to get along with other humans and God, using a liberal arts curriculum. That is what I needed in my journey to wholeness. All the disciplines are now integrated that I have learned so far. I like that. It feels good. I have come to see that I have a lot more to learn, but I feel I have the big picture now to guide me. One of the great lessons I have come to see is that I need to grow and be nurtured in the disciplines that count towards the big reward for building the character that fits the purpose of my existence of my Highest Power, the Creator God. George Armerding on page 5 of his book, The Fragrance of the Lord, gives us an example of how the disciplines can blend in his analogy of fire to trials that are necessary sometimes to produce a sweet spiritual smell in us: "Not every odor greets us as we simply pass by. Many delightful smells are...in the heart of a tree, in the kernel of a nut, in the root of a bush, or in a leaf that must be crushed before we can enjoy its fragrance. Sometimes one may have to resort to fire to persuade nature to release an exotic odor, for even the word perfume means "through fire and smoke" just as the word incense means "in fire." Only such torture will yield a quality of delight that otherwise would remain hidden forever. Thus, those whose lives have been subjected to extreme pressure or fiery trials often yield the sweetest fragrance." I do not like to go through my trials except to know that they work together for good. They bring forth a peace knowing I have survived and have now learned to give the sweet odors of compassion, comfort, and consoling on anyone else going through a similar trial. I am smelling better all the time after each trial if I have gone __through__ the problem dealing with all of the heat no matter how hot it gets! I receive support from friends and my Highest Power in order to overcome the problem and obtain victory! Victory is another sweet odor I delight in but cannot be obtained if I do not go __through__ the problem and/or heat!

And presently, in 1991, I am studying in a few "support groups" of anykind in order to get a much better understanding of the interaction of people in groups. It is in the "support groups" that I can learn just how human we all are. There I learn about the family systems with their blended functions and dysfunctions that have such a large influence on our lives. I learn there that there seems to be no problem that is not common to other individuals. For example, I can bring up the toxic shame I was given as a young boy and teenager about my breath problem and how it affected me. I have always thought that I was probably the only one on earth so shamed by this problem, that I was less than other people in value because of it; but the "support group" individuals will give their examples of this problem also if they have had it or have it. This can be done during the meeting or during a break. There are alot of different kinds of "support groups" for almost any problem you have; for example, alcoholic, sex, love addictions, overeating, drugs, codependency, gambling, etc. I like the non-shaming environment of the "support group" where I can discuss nearly everything without toxic shame. The support group" allows me to learn to talk about _personal_ problems without being toxically shamed. It allows me to see how I am on a journey of being human; that I also am a human being with the spirit of man just like they are; and that I also need help to grow and be nurtured with the help of other humans. It also shows me that man needs a Highest Power, God, to guide and direct him beyond the spirit of man in order to see the purpose that He has for His creatures made in his image, man.

Thus this book will be an interdisciplinary study of the whole man in science, health, liberal arts, "support groups", and the Highest Power including man's dependent, independent, and interdependent relationships as they relate to the sense of smell. To me there has never been a greater need in human history to find out answers to the human problems and questions dealing with this subject. When I first started writing on this subject I noticed the scarcity of material compared to other subjects such as exercise, health foods, medical books, other types of recovery books,etc. I needed lots of material from books to write this book in an interdisciplinary manner, but I found them few in number todate. I asked people to write testimonials dealing with odors, but many told me that it was a "taboo" to write about this subject. Some said it was culturally a greater "taboo" to write on this subject than sex itself as a subject. Steve Van Toller and George H. Dodd write on page 4: "To refer to another's body odour in public-or indeed one's own body odour-is socially unacceptable; it is a behavior which civilization has outgrown and outlawed." (see the Bibliography whenever names are given throughout this book) I have found a few brave individuals who are willing to write testimonials, and I hope that there are more of you out there who will fill out the testimonial sheet and sign it and send it to me as soon as possible so that I will have it for my next edition of this book.

An example of a testimonial given by a man who writes on the cultural aspect of odors is as follows: "The Olfactory: Its Effect On Cultural Conditioning(his title) What odors are pleasant and what odors are unpleasant? As a part of the study of sensorics as found in the study of Intercultural Communication, the olfactory can shape our perceptions of other people and places in either a positive or negative way. It is important to remember that so much of our positive or negative perceptions relating to our sense of smell is culture Bound. It is evident that cultural conditioning can and does effect our perception of odors. To illustrate this point, the Highlanders of New Guinea bath themselves with mud and pig grease and rarely take a bath. Within this culture, the resulting odors are considered acceptable. However, this odor could be contrasted with the meticulous Ghanaian who may take several baths a day (Dodd, Carely H., _Dynamics of Intercultural Communication_,pg 191, 1987). The difference of these contrasting odors would be easily discernable to the unconditioned nose of a person outside his/her own culture. The North American culture, in general, has very strict rules about unpleasant body odors. This culture has spawned a proliferation of deodorants, and a host of other ways to cover up body odor. The body has *a number of ways to rid itself of unwanted toxins. One major way is through perspiration. In some individuals these toxins are strong enough to produce a powerful body odor. Another way toxins are removed from the body is through the lungs. These toxins may also effect a persons breath odor. Television commercials frequently advertise breath deodorants to combat 'bad breath' produced by these toxins being exhaled from the body. Within the realm of pleasant and unpleasant body odors, there is a sense of appropriateness that should be taken into consideration within a particular culture. For example, the North American culture, in general, finds that even unpleasant body odors are acceptable in certain situations. For example, the smell of perspiration is considered acceptable in a locker room. However, this same odor would be totally unacceptable in a different situation such as a social event where there would be a formal gathering of people. Cultural norms can and do determine when and where certain odors are acceptable. In any case, perceptions of odors whether positive or negative are largely determined by cultural conditioning. Signed Albert Bruhn."*

Like this testimonial, your testimonials may help many people. They can be short or long, any length. I found that testimonials are important in interdisciplinary studies as they show me and others that we are all human having common problems that we are struggling to overcome victoriously. Testimonials, the word has in it the word "test", indicates that something has been put to the test and has worked for someone, and hopefully it will work for others also. So please send your testimonials today. Also, my next book, of which I hope to have out in 1992 or 1993 will be on the subject of snoring. So if you have any testimonials on snoring please send them to me now using the same form in the front of the book and changing the subject to snoring, or else write for the snoring testimonial sheet. I want to thank you all for your participation in advance. It is ok if you do not participate in testimonials; although, I do hope you will pass this book on to friends so that shame over the subject of odors can be eliminated. So now please read on to find out just how the subject of odors from head to toe can be addressed in recovery and in ecstasy! I feel hopeful that the information in this book will be profitable for your growth and nurturing on your journey to wholeness!

MOUTH RECOVERY!

Bad breath can cause misery to the person possessing it as well as to those nearby. The good news is that it can be conquered in most cases! There are a number of things that can be done--never fear that it is impossible to conquer or at least lessen greatly!

Oh how I know that many of my readers have suffered, or are right now suffering, from this horrifying problem! It is so horrifying to open your mouth for fear that a mate, friend, or relative might be struck negatively by the terrible odor that you hate down deeply and wish with weeping sometimes that you never had. The pain is sometimes so excruciating mentally and emotionally as you sit or stand next to someone with anxious desire to speak, but you know the sorrow that it will bring if you just opened your mouth ever so slightly letting even a little bit of that awful smell out. This is because it might offend someone you love or care about. The fear of the unknown reaction to this smell is frightening and paralyzing to you; so communication stops until you are once again far enough away from anyone who might breath in your bad breath called halitosis.

But let's get one thing straight from the beginning: <u>You</u> are not <u>the cause</u> of your bad breath! THAT'S GOOD NEWS! Something else is the cause of that awful effect. There is a cause for every effect. If we eliminate or minimize the cause, we eliminate or minimize the effect. THAT WILL RESULT! Thus either no more bad breath or a minimum amount that only those very few extremely odor sensitive people will detect. But the good news is that there is even covering up of this "minimum amount," just like covering up blemishes with makeup for the face. This book will go into that process in the next chapter.

And one more thing! It is not a shame to you personally for having bad breath. You have done the best you knew how to before reading this book to stop it. In fact, that is a glory to your credit that you creatively worked to find an answer for months or years to fight this problem! This has given you your present overcoming attitude that has thankfully now matured and can be used with the material in this book. What a powerful combination you now possess that, if put into faithful practice, can lead to successful victory in overcoming bad breath and replacing it with <u>good</u> breath!

<u>*The Causes and Removers of*</u>

Bad Breath

Now let's discuss the causes of bad breath. But before I do, remember that professional care such as from a Medical Doctor is <u>always</u> important in <u>any</u> heath care matter. This is not just to treat the problem, but also to see if you have any allergies or other heath related problems before beginning any suggestions in any book you read. Also, the professionals may help you if you have an addiction, for example to cigarettes, that can be a source or cause of bad breath. And remember that the directions stated in this book are in no way to be considered as a substitute for consultation with a duly licensed Doctor.

Now these are the causes and removers of the effect called "bad breath" or halitosis: (Please write to me anytime using the testimonial form at the front of this book if you can add to this list. Your comments probably will be used in the next edition. Send them to my Colorado address shown on the form. Please do it right away. My thanks go out to each of you that do help by reaching out to the many people suffering with this problem!).

1. *Dirty teeth need cleansing of teeth.*

2. *Dirty cheeks need cleansing of cheeks.*

3. *Dirt between the teeth need cleansing and flossing.*

4. *Dirty hard palate needs cleansing of the hard palate.*

5. *Dirty tongue needs cleansing of tongue.*

6. *Dirty hidden areas like between the cheeks and gums needs cleansing.*

7. *Dirty gums need cleansing and professional care to remove tartar at gum line and under the gums.*

8. *Dirty bacteria and/or yeasts that enter mouth from outside sources that need professional care or advice.*

9. *Rotting teeth need professional care.*

10. *Lack of nerve supply to mouth area needs professional Chiropractic care and/or other professionals.*

11. *Stressful situations can aggravate condition which needs peaceful environment, following peaceful principles, and more sleep.*

12. *A smell that someone does not like that another person may enjoy as a personal preference. This would require changing a personal habit around those individuals. For example, some people do not like the smell of a certain kind of toothpaste or food eaten.*

13. *Also, some people do not like the odor of cigarettes or cigar breath.*

14. *Intestinal problems need professional care and counsel.*

A Personal History as an

Example of Recovery

I will now put all of these together so you may be encouraged in your recovery or in the recovery of another person. I am a recovering and recovered (recovery is nearly 100% complete) bad breath victim. This is true in that as in all recoveries there is a cause for every effect; and therefore, the ways to recovery must be <u>maintained</u> even after recovery in order for no backsliding to occur! I also found that the cleaner and purer a life I live, the more I am blessed in overcoming any problem. I found that I need to replace all bad causes that produce bad effects with good causes that produce good effects. If all I did was coverup my bad breath with, for example, a mouthwash, without removing the cause such as bad gums, then the problem simply returned. Sometimes the mouthwash only mixed with the bad breath. But a good heathy mouthwash with the cause of the bad breath removed is wonderful!!

I now want to tell you my story. I had the problem of bad breath from the earliest years of my life that I can remember. The torture of watching faces make that awful grimace right before my very eyes with nearly each person I would talk to was so painful. I remember feeling shamed by their expressions on their faces, that I was less than the dignity God created me with. There was a feeling of <u>worth-less</u>-ness, worth less than the value I really have before my Creator who made me in His image! The good news is that it really was not I that was <u>worth</u> <u>less</u> than the dignity God gave me; but rather, it was only the bad causers that were causing me to think that. This is true since once these bad causers were gone, the problem <u>left</u> me! I did not leave, the problem did! I was a person who was the victim of these bad causers such as offender bacteria that were rotting my mouth as an attack on <u>me</u>! But I am here to say that I am a survivor! But more than a survivor, I am one who now tries to overcome all problems by striving to only do the principles that will magnify the Creator and His dignity He has given to all peoples. His ways work, I just have to work them! I am in His image; therefore, I must do the principles He does in order to be like Him. In Him I must move and breath and have my BEING!

I want to never go back to the time when a student of mine in my Bacteriology lab class responded to my bad breath after he asked me a question: "It was not worth the answer!" This he told me about one year later into my recovery: this was about 1971 when I was about 24 years of age. I also never want to go back to the time when I would ask girls for dates, and they would make such awful faces at me! Other awful experiences hopefully never come back to me: all I need do is remember *that there is a cause for every effect! A good cause brings a good effect; whereas, a bad cause brings a bad effect!*

So my journey to wholeness has been a hard one, like many of you reading this book, but the lessons along the way and the results make it all worth while! For example, I had no one to turn to who had the answers so complete as I had to learn little by little along the way. That is why a good support group would be good for those suffering from any odors that bring a feeling of shame, but unfortunately there was none then and perhaps not even any now. The good news though is that this book should help many of you suffering from this problem; and perhaps, a group could be formed in a proper manner in your local city. Now please let me share with you what I have learned in my journey over 40 years!

I was born in Culver City, California in 1947 with health that was not so good. I remember that my parents said that I had colic from the beginning as a baby which is an uncomfortable intestinal problem that makes babies cry in pain.

Intestinal Problem:

That intestinal problem got worse but changed into a chronic constipation that remained that way until I was about 23 years old. Chronic constipation does not allow the food to move out of the intestinal tract quick enough to avoid its putrefying into smelly chemicals and gases. This I was later to learn was one of the many causes of my bad breath; because, the bad bacteria in the intestines produce bad smelling (the effect) gases that go into the blood where they are carried to the lungs to be breathed out when I would talk. I tried all kinds of diets and nutrition, and even went to the Medical Doctor and the Chiropractor, that would work for other people to get rid of my constipation; unfortunately, none worked for me. But rater, this seeming unfortunate situation was made a fortunate one for me; because my life became unmanageable, I tried all I knew to do, I could not handle the situation, so I turned it over to the One greater than I who did a miracle for me. After praying to God to heal me through a Minister, I was no longer constipated! Later, in 1989, I was to learn that an improper intake of vitamin C can be a major contributing factor to constipation.

Let's get back now to the bad bacteria that are a cause of the bad breath. Another way to get rid of the bad bacteria, besides the making of the intestines or bowels move, is to eat foods that favor the good bacteria and eliminate or decrease the growth of the bad bacteria. So I learned to read about which foods to eat that would reestablish the good bacteria to replace the bad smelly gas producing ones. I read that garlic was good for this and it worked well when taken for months. But I soon found only my garlic loving friends could tolerate me. This I found puzzling since they actually liked the smell of garlic on garlic bread and in an Italian restaurant but not on me or on themselves after they left the restaurant. I still like the smell as it reminds me of being in an Italian restaurant with my family in my earlier days between the ages of 7 years and 20 years when some fond memories are made. Get this: we used to have conflict over the garlic bread, who could have the most; but later stand about 3 to 4 feet away from each other so as not to smell the garlic on each other's breath. Does that make sense to you? Later I was to learn a trick about garlic from an old lady who came to my office. She told me to cut the garlic, even taking off the peeling, under water; then she said to chop it up under the water into small pill sizes and then swallowing them. This should only be done if you can swallow pills easily and only a few pieces, not the whole thing, at a time. There was no garlic breath from that as apparently it takes the air's oxygen to make the smell. I later found that some companies sell odorless garlic by the name of Kyolic Garlic in the Health Food stores.

One person wrote this about garlic: "I would like to come to the defense of garlic! This is a wonderful food that has many healing properties and shouldn't be ignored because of it's smell. Pavlo Aerola

has written an informative book on garlic that you can probably get in

your local health food store.

Jenai Rasmussen 5/14/91"

Also, I learned to put back the "friendly" or good bacteria that do not

make the "smelly" gases. These bacteria are called Acidophilus bacteria.

I learned that these bacteria can be lessened in numbers if we eat foods

that do not feed them or even injure them, take enemas or colonics too

frequently without replacing them, or take certain medicines (this would

be a good example to remain under a Medical Doctor's supervision to

monitor your friendly bacterial population). I found some companies sell

Acidophilus bacteria in Health Food stores. I still take Acidophilus

bacteria even to this day as they not only are supposed to balance off the

bad bacteria but also they are supposed to make certain vitamins for us.

So check with your Health Food store and professional on how you

personally need to take it if this is one of the causes of your bad breath.

Remember that these good bacteria are naturally found in healthy people

in goodly numbers to fight off the bad odor causing bacteria that befoul

your breath.

Further, I learned to eat more fruits and vegetables and to stay away from any food that I felt made my breath undesirable. For me personally, this included milk and milk products, chocolates, and fried foods. I also found that a simple allergy test helped me to determine which foods my body found rejectionable; and therefore probably a source of decay in the intestines. It helped my breath, it may help yours: You can check with your Medical Doctor, and some Chiropractors are reported to send out patients for allergy testing. Also, if the food is not properly digested, a good digestive aid may help like the ones I buy from a vitamin company; so that the food is not incompletely digested which leads to putrefying and bad gas formation. This leads me to another point I learned in my journey to wholeness in living a whole and healthy life. There are small unnatural pockets that bulge in some people's intestines that collect old food that rots in these garbage can like pockets that need to be emptied, each called a diverticula. I used psyllium seeds that were processed correctly to act like a broom to sweep out the intestines and clean out hopefully all these bad bacteria and their food that they make bad gas out of in these little unnatural sacks or pockets. This technique is a good one that should be under the guidance of a health professional.

Before leaving the discussion of the intestines, I feel it would not be complete without mentioning intestinal tissue integrity or the strength and health of the intestines themselves to move the waste material out and to provide a good environment for good bacteria to outnumber the bad ones. Taking some good vitamin and mineral supplements can do this, keep the tissues of the intestines strong so that the walls of the intestines do not get sick and serve as food for bad bacteria. Chiropractic manual manipulation of the spine to keep good government nerve supply to the intestinal tract, as well as the mouth, can also serve to keep these parts healthier. I still have regular Chiropractic care and take my needed vitamins and minerals that seem to definitely help in the overall picture to make my mouth smell better. This is probably because it keeps my intestines and mouth's gums and other tissues stronger so that bad bacteria cannot attack them. And the Chiropractic care keeps the nerve or government supply strong to direct the vitamins and minerals to where they are needed. Can the intestines move correctly at the right time without the nerve or government directing it in coordination with the rest of the needs of the body?: of course not! This can be a source of constipation where the bowels or intestines do not move the fecal matter out quickly enough which can lead to foul smelling gases!

<u>Cleansing and Healing the Mouth for Better Breath</u>

This brings me to a discovery that helped to change my life, and that hopefully will change yours also, in my journey to wholeness. It was on a skiing trip with friends from California in 1984, if my memory serves me correctly. My breath by that time was improving pretty good as long as I did not get extra close to people: about 2 feet away was very safe at that time. I remember when in the 1960s and 1970s some people told me they could smell my bad breath across a table; one person joked, which was not funny to me at the time, "Across the whole room!". It is interesting to note here that not all people have the same power of smell. For example, once my dear dad who died in 1975 only 6 months before I met my wife, asked me why I do not go on many dates like my two brothers. I told him, "because my breath smells." He replied that he could not even smell it bad at all. This reply was on a day that I just received word from someone that a girl told them that my breath stinks. My dear dad had been smoking for years and that is probably the cause for his inability to detect the foul odor of his precious son, Ronny! And Robert Burton on page 109 of his book gives another possible reason: "...some people are insensitive to smells as the color-blind person is to colors. There is also a variation in sensitivity which is related to the anatomy of the nose. In some people, the shape of the nasal passages hinders the passage of odour-laden air into the nasal clefts so that they may be unaware of *odours unless they seek them by sniffing hard." But even that worked together for good since I, although cautious, was able to be close to him when I was talking. This is in contrast to my wife's very, very, very, sensitive power of smell. Robert Burton in his book, The Language of Smell, on page 110 gives a possible reason for this, "...the simple and well known observation that women are better than men at smelling certain substances..careful experiments do show that women detect a range of substances at far lower concentrations than can men." She is my ultimate test to see when I have fully 100% recovered! God gave her to me for many reasons to be my wife including this one! God is fully of knowledge, understanding, and wisdom in all things to bring us to the place where we will someday know we are whole!*

So back to the skiing trip! I saw 4 of my friends with tooth brushes in their hands, males and females. I curiously asked them what they were doing in carrying around their toothbrushes in their mouths and pressing them against their gums for literally hours as they walked within the condominium. They told me in great excitement, "We are blotting!". I asked them what that was and why would they want to do that so fanatically! They said that it keeps their gums and mouth in such good shape that their breath stays as pure, "as a baby's breath." WOW! That sounded great to me, and too hard to believe. Guess what? It worked great as long as I would do it daily. All kinds of tartar and white stuff came out on the tooth brush that I had to keep cleaning off before inserting it again and again! I now could get about 1 1/2 feet next to anyone I wanted to. Tears almost came to my eyes as I saw nobody backing up away from me and making those terrible faces because of my _former_ horrible breath that now was only _slightly_ foul! Confidence began to come to me! My personality began to blossom! But I had to keep this up everyday! The days I stopped doing this, the odor slowly would come back; I have learned my lesson now, I rarely miss a day! This technique should also, of course, be done only under the supervision of your Dentist before and throughout your doing it. To do it, just get a soft bristle tooth brush and press it gently against the gums for minutes in different places, not staying in any one spot. My Dentist told me that my gums should be carefully pressed and not to press if the gums bleed. So I pressed keeping away from the areas that tended to bleed, and eventually I noticed that the healing was spreading so that there was no more bleeding anywhere. My Dentist was surprised at the quick results for me as he said my gums were so bad. My teeth even looked a bit cleaner. And best of all, my breath was not so bad, even a sweet smell began to emerge!! The gums that once gave off a stinky smell were now healing up. I liked this good cause (blotting) for this good effect (better smelling mouth and healthier gums). I remember telling myself that I really was not sick, there was nothing wrong with me, the inward man, only the outward body was sick! Also, there was no more horrible smell from the bleeding gums. This was great for me in my journey to wholeness where all of the whole body, as parts working together, worked coordinately and healthfully! I hope this helps many of you!

I need to mention flossing at this time. One dear friend of our family told me, "You only have to floss the teeth you wish to keep!" Of course I wanted to keep them all! What was the point? My Dentist told me he found that flossing strengthens the gums which support the teeth, and that the gums do not break down and give off a smell with good flossing technique. Flossing also clears old smelly decaying bad bacteria filled food between the teeth where it gets stuck and putrefies: the longer it has been there the more it stinks up your breath. There is an art to flossing that your Dentist can teach you. My Dentist needed to show me how due to the condition of my gums that I did not want to injure in their weakened condition. My Dentist said to take the dental floss sold in the store and move it carefully between all teeth cleaning every side of every tooth and also massaging gently the gums with the flossing string.

Which brings me to a precious discovery: massage! Under a professional's supervision, this is a wonderful experience as well as a strengthener of the tissues so that they do not get sick and thus begin to smell befouled. Make sure your doctor ok's massaging your gums and checks and upper and lower tissues behind the upper and lower lips, respectively. Also, a professional will show you the right way to do this so that wrinkles will be avoided. I have learned and profited much in terms of having sweeter smelling breath by doing this. The fingers, when very clean, can be placed in your own mouth and gently used to massage the tissues to bring more blood circulating to the areas for healing. Note that if you are taking good vitamins and minerals that this will better bring these to the area for pick up and use in the tissues massaged. And the Chiropractic adjustments of the vertebrae will allow the nerves to send governing messages to the tissues to best use the minerals and vitamins.

Another great blessing has been to learn about tongue brushing. My Dentist told me to gently brush away the white off of my tongue. This has helped my breath very much. I can smell the difference before and after brushing. This brings me to a point that my dad taught me in the 1950s: he told me to wash my mouth out and clean it every time I would wake up at night to go to the bathroom. The breath really smells the worst usually in the middle of the night and when we first wake up in the morning; but a healthy rinsing first of the mouth with water, followed by a good teeth and tongue brushing does great wonders! Also, using the clean index finger in the mouth in those hidden areas adds an even cleaner feeling and smell to the mouth. I do that one probably every night, morning, and throughout the day for a more clean mouth smelling me. Also, ask your Doctor how to get rid of the old waste material in the back of your throat; I get rid of mine, and it is hard to believe what smelly stuff the air from the lungs has to travel over before I talk to someone if I do not get rid of it! So now I not only clean my teeth by brushing them, I also clean between them by flossing, and strengthening their support system the gums, by massage and cleaning with my finger.

I also clean the hard palate with my thumb gently the same way back and forth which removes any smell causing bacteria that are lodged in between the folds of tissue there. This I do every time I feel my breath beginning to smell, and after it is done, my confidence returns instantly. This is done on the cheeks too on the inside of the mouth.

There is also a new blessing for all of us in the form of home cleaning the teeth. I use the one that has three parts to it because I like the second *one that has hydrogen peroxide in it. The hydrogen peroxide is known for its killing of bad bacteria in the mouth; which in turn, stops rotting of food at the speed it increases at due to the bacteria which cause the bad smell. Check with your Dentist if you can use these home cleaning methods and how to use them. I noticed that my teeth are so white compared to before that even the very tiny areas for bad bacteria to grow are even cleaned out. This I learned in 1991 which shows me that my journey is a process and not a destination.*

Another important aspect of keeping my mouth clean has been mouth washes. The two that have helped me the most are chlorophyll and water, and sea salt and water. I dissolve either chlorophyll or sea salt in water and then rinse my mouth out. One person commented about the chlorophyll: "Chlorophyll has miraculous properties in addition to purifying your breath, it also serves to purify your blood. It comes in both a liquid and tablet form and can be one of your best health aids! The Rolls Royce of chlorophyll is sun chlorella of which is manufactured by YSK International Corporation. Signed, Jenai Rasmussen."

Also needed to be mentioned is my every 6 month dental cleaning that is done to remove the hidden smelly plaque which is not only on my teeth, but also between my teeth and gums down deep. Many times this can be the smelliest mess if it has not been cleaned out in months. I know of no other way to get rid of this except by the Dentist, there is no way I can get to this no matter how much I clean my own teeth since it is hidden. This is another of those unmanageable areas in life that can get much out of control without help from someone with more power (in this case the Dentist) than myself. If I avoid the Dentist, this stuff will putrefy more and more and smell worse and worse!

Also the Dentist should check to see if your dentures are being cleaned correctly if you have them. Dentures can hid bad bacteria and wasting food particles that smell bad. Follow your Dentist's advice and clean those dentures regularly.

The Dentist can also take care of rotting teeth which smell terrible. When my teeth that were sick were helped by my Dentist, the smell of my breath improved noticeably by myself and others. It is interesting to note here how so many things we can do for our breath can all add up accumulatively; the more we work these the more they work for us.

Another thing I did was to stop eating foods that smelled good to some people but made others tell me, "Your breath smells!". This is personal preference as to what smells good and what does not. Some people do not like the smell of chocolate or cookies or milk or garlic, others like these smells. So I learned to eat only certain foods around certain people who are special to me like my wife. I also work hard not to offend others. Please note that the reverse is also true; some people love your breath when you eat a certain food, I will make notes on what special people like concerning my breath and eat those around them.

Another problem that needs to be looked at is bad breath due to throat infections and lung infections that a good health professional can help with. Sometimes just leaving the window open a very little bit, as long as no chilling results, can help keep the bad bacteria down in numbers since they are usually anaerobic (they grow best when there is no air of oxygen). The fresh air has oxygen in it that helps keep the bad breath producing bacterial count down.

Another thought that I would like to leave all my readers with that may be a cause of bad breath and was worth a try for me is the removal of the mercury-silver fillings in my teeth and replaced with a "composite". This I did in June of 1991 in Longview, Texas. I got a big surprise when I did this. My gums now look more pink than ever and my breath appears to be much improved. The question in my mind as I write this today on June 26, 1991, is did these fillings have an allergic reaction in my mouth when I first got them about the age of 7 years that caused my gums and other structures in the mouth and body to smell unpleasant? I had heard that their are some Dentists that think the mercury can cause some side effects. Whether this is true or not, the results are very encouraging to me! I do not find that my mouth odor is as bad in the morning either after these "composites" replaced the mercury fillings. I told my wife that I felt that my breath was better and she replied, "You're breath does smell better." Those words on my journey to wholeness are wonderful! Also, the better looking teeth are also worth boasting about since the ugly mercury-silver fillings are now gone, and my teeth look more like the way the Creator made them!

So go for it if you are excited to make your journey to wholeness more complete. You can have a sweeter breath and thus sweeter friends with faces that smile and not grimaces due to bad breath from the mouth. See, you are ok, the problem has been removed. Your being is ok! Next I need tell you my story of the use of good products to make the breath better or even exciting!

ECSTASY OF MOUTH ODORS

It feels good to know that the mouth smells great; that the people you are near can actually enjoy your breath with every word that comes out of your mouth. The look of pleasure and curiosity on their faces is priceless and makes a good memory impression that you can return to nearly anytime in your life! I have had people tell me in their delight, "Wow, what have you just eaten, you smell good!" This is like a dream come true for a person like myself who had for so many years been shamed by family, friends, and neighbors of all kinds because of my bad breath. I really like to put my emphasis on ecstasy much better than emphasis on recovery; although, both are necessary! What I have learned dealing with mouth ecstasy is that different odors are ecstatic to different people; and therefore, once again, the individuality of the people we are breathing on must be considered. I feel it is wise to make a list of my favorite friends that I want to have different ecstatic results with!

I also have had a better and quicker recovery from my being shamed, where I was made to feel that I am a disappointment, by now surprising those same shaming people with delightful breath! This I know, that odors of ecstasy can be used by anyone that has or has had such a shaming experience!! The recovery is to remove the disappointing factor, *for me the disappointment to those near me every time I opened my bad breath mouth, and replace it with something better, something ecstatic that gives joy to the person(s) that formerly were shaming me through toxic grieving (complaining). To me the word, recovery, is an exciting word in that I divide it into two parts: "re" means again, and "cover" in covery means to cover. This tells me I must put a cover either by eliminating or stopping the expression of the problem, in my case bad breath, every time it tries to get uncovered and effect my life in a negative fashion. This goes for recovery in anything such as the <u>temptation</u> to get drunk or do a love or sex addiction; when the thoughts that are already present from the shaming words and being that binds us, raise their "ugly head" in our minds, recovery is used to just cover them again and again through words and actions that have the <u>power</u> to cover them. These temptations must be looked upon as only being what they are: TEMPTATIONS!! The right power can cover any temptation!!!*

For example, if I get tempted by a thought or the appearance of a milk product that will surely make my breath smell, I must avoid the bondage that thought or appearance wants to bring me into by using a higher power, a greater power, than that of the strong <u>hold</u> of the temptation. I like to use various "strong hold" breaking techniques using various powers that are greater in magnitude than the power of the temptation. This is called overcoming! I overcome the temptation that has been imprinted in my mind either from my family of origin, or my society upbringing, _or from some other source_. For example, if the temptation is there, which it will be, I can think about the consequences of eating the milk product: people's faces in agony and shaming me! That will stop me, but not always! If I think more positively, like the happy and joyous time I will have if I do not have the milk product, then I will usually stop and re—cover the temptation (like with a blanket or lid). If I think and visualize this positive for a longer length of time, the results of re—covery are even more powerful. If I think of what character I can build by re—covering many times by habitually covering or overcoming the temptation, and I enjoy thinking of the rewards in my relationships it will have, then that gives me even a higher or greater power over this _seeming_ strong hold! I believe with every bit of reality that there is always a _good_ power to overcome the strong hold of a _bad_ power that brings the unhappiness and suffering to our lives. AND THAT'S GOOD NEWS! If the highest power of good can be exercised, then even the highest power of bad can be overcome! I also found that the use of the highest power I know of has helped me to overcome many, many, many temptations better and quicker *than high powers. I hope to write more on this later or in a future book!*

Before I move on with the subject of ecstasy of mouth odors I want to give my opinion on codependency in this matter that may help the reader in overcoming and re—covery. The word "co" means with or together; and the word "dependency" means to depend on. In our example, if I was dependent or bonded to my feelings of satisfaction when I would eat a milk product, and those feelings were highly desired to the degree that I would sacrifice my health and feelings of well-being and victory in overcoming; then I would be hooked or bonded (in bondage) or dependent on fulfilling the temptation in order to achieve the feelings, not of well-being and victory which cannot here be obtained, of fullness and satisfaction. Anybody, such as a wife, who helps me in my addiction or bondage is a person who is "co-" or with me in my dependency. For example, the person may want to please me so that they will give me my milk product. Unfortunately for the one who is codependent, they must suffer for doing this codependency; in this case, whoever gives me my milk product will get back from me a blast of bad breath which could hurt our relationship with one another. Better would be if the person would be strong acting like a higher power to help the re—covery of their spouse or friend by being a good model of how to resist temptation by a re—covery process of their own problem; and to also model how to resist temptation in any food that may be cause themselves bad breath! If we all model good for each other, we will grow in the qualities that build our own self esteem (our own real value as an individual in our identity) and thus be a higher power for overcoming our own temptations that we can overpower. Of course, there are always some temptations that only the highest power, God as our Creator, can overcome! I have come to see how limited I am! I am a human with the power of the spirit of man that the Creator God gave me to understand the things that go on in my human life; blessedly, I found that the Spirit of God is a higher power than the spirit of man which can conquer and overcome in victory any temptation. The ways of the higher powers all work, I just have to work them; and my greatest joy is in working the paths of the Highest Power, God the Creator who loves us all unconditionally. I desire greatly to show how the different powers work in my life later in this book or in a future book. The Highest Power led and started me in my journey to wholeness to recovery in 1969. I have to respect Him and His wonderful works in me as I journey through recovery toward an every intensifying life, in connection with Him, of ecstasy!!!

Now more on ecstasy of mouth odors:

What mouth odors do you enjoy the most? Stop and think of which ones you enjoy smelling on others; but most important of all, stop and feel and think about which odors you enjoy and delight in the most! Think and feel about what odors you can be influenced to help make your day a better day in your recovery. Which odors can give you a <u>confidence</u> in your talking to others? This is helpful to me, for example, since the shame tapes in my mind of how awful my mouth smells can be put to temporary silence by overcoming them with the new tapes of how glorious (glory is the opposite of shame since shame is what we want to hide and not expose, but glory is what we want to give to others and show them) I do now smell. I can now choose to listen to whichever tape I want. I can have that much control of my life as long as I keep adding truth to my new tapes and practicing what is on them to form a habit or character of who and what <u>I am</u>. And it is that Iamness that is now what I value which becomes my true or real self-esteem; not the false self-esteem based on externals like my money value or my position in society value since all of these come and go or change too often. So the mouth odors which change according to what we do externally need to be used to help us in our recovery and beyond into ecstasy.

One odor, if you like it, that can be used is cinnamon. I personally like that smell on my breath and on others. In his book, Heinerman's Encyclopedia of Fruits, Vegetables and Herbs, John Heinerman who, " is a medical anthropologist whose research has taken him all around the world...," has this to say about cinnamon: "Fantastic Mouth Wash. In place of Listerine try another antiseptic mouthwash that really does 'kill germs on contact.' Half a teaspoonful of tincture of cinnamon added to half a tumbler of warm water makes an excellent mouth wash when the breath is unpleasant and the teeth decayed. To make a tincture, combine 10 1/2 tbsp. powdered cinnamon in 1-1/4 cups of vodka. Add enough water to make a 50% alcohol solution. Put in a bottle and let set somewhere for two weeks, shaking once in the morning and again in the evening. Then strain and pour the liquid into a bottle suitable for storage. This tincture will last a long time (page 99)." He also says on the same page, "...true cinnamon is more desirable in sweet dishes, pastries, breads and cakes. Cinnamon was included as a major ingredient in a 'holy anointing oil' that Moses used." Just the smell of cinnamon on a bread or sweet dish makes my mouth water. I have seen the same reaction in others including little children. The joy in all this is that it will make the breath smell like cinnamon which may attract in a correct way the right person to you. The recovery issue here is that it triggers off the right tapes in the mind of character (what you are) growth and not those that will put you into temptation. If the wrong tapes are triggered off, the power to cover the temptation, to overcome the thoughts of temptation must be used as shown in an earlier example. By the way, John Heinerman's book may be purchased from the Parker Publishing company in West Nyack, New York 10995 (you can ask the information operator for the phone number to order it over the phone, probably). Remember to always check with your Medical Doctor before doing anything in this book incase of allergies, etc.

Other odors that not only heal but smell great to some people that John Heinerman writes about are carrots and watercress, cloves, coriander, dates, dill, filberts, parsley, peppermint, wintergreen, and thyme. He also talks of myrrh and goldenseal as healers. Let's look at some of these smells that you or a loved one or friend may enjoy or delight in:

"One of the most popular remedies among chinese residing in Hong Kong and Canton, china is a special watercress soup used to treat...bad breath. There is no specific amounts called for, but generally for one person, about 1/2 lb. each of cut watercress and chopped carrots are cooked in 2 qts. of water. The liquid is boiled down slowly to 1/3 or 1/4 of the original fluid volume and then the soup is consumed with the *vegetables intact (page 370)."*

"Stops...Bad Breath. Clove is a powerful, penetrating antiseptic which makes it ideal for an effective mouth wash. In 2 cups of hot water, put 3 whole cloves or 1/4 tsp. ground cloves, and steep for 20 minutes, stirring occasionally. Then pour through a fine strainer and use as a mouth rinse and gargle twice a day for bad breath (page 108)." I have personally used cloves bought in the Health Food store or market that I will suck on or else bite into that some people seem to like from the expressions on their faces.

"Eliminates...Bad Breath. In the southeastern mainland China city of Canton, coriander leaves and seeds are used...to get rid of halitosis or bad breath. Bring 2 quarts of water to a rolling boil. Reduce heat and add 3-1/2 tbsp. of seed. Simmer for 1-1/2 hours until the amount has been reduced to slightly less than a quart of liquid. At this point, add 2 tsp. fresh, finely grated orange peel and one pitted, finely chopped date. Simmer for an additional 15 minutes, at which time remove from heat entirely. Add 1 tsp. each of dried coriander leaf (if available) and finely chopped fresh parsley, with a drop or two of peppermint oil or wintergreen oil (if available, but not necessary). Steep mixture for about half an hour, stirring occasionally. Strain through a fine sieve or filter

paper and store in a pint fruit jar with a good lid to seal it. Store in refrigerator until needed...gargle and rinse mouth with 1/2 cup of it while cool, but not heated (pages 113-114)." I have personally used the seeds that I bought in a Health Food store. I personally find the coriander seed is odor eliminating with a nice gentle comfortable smell as a lingering odor.

When I chew on these tiny seeds, very few at a time, I once took too many and it tasted rejectionalbe, I find my confidence really skyrockets as my breath has a delicate good smell of coriander that people seem to enjoy. In fact I had some today before I came to this computer lab to write more of my book in which I asked questions to people with assurance! Today is June 11, 1991.

"DILL...Gets Rid of Bad Breath...Try chewing some <u>dill seeds</u> the next time you experience halitosis. You'll be pleasantly surprised to see how quickly they sweeten and freshen your breath (page 125)."

"Sweeten Your Breath with Filberts...The breath may be somewhat sweetened by chewing on a few <u>filberts or hazelnuts</u> for awhile. They work not so much because they're aromatic like peppermint would be, as they do by absorbing much of the bad breath like a sponge does water. They're quite expensive, but worth investing in for this purpose...(page

233)."

"<u>Myrrh</u>...Cures Breath...In 1 pint of boiling water, steep 2 sprigs of coarsely chopped <u>parsley</u>, 3 whole spice <u>cloves</u>, 1 tsp. of powdered myrrh and 1/4 tsp. powdered <u>goldenseal</u>. Stir occasionally while cooling, and then pour the clear liquid part through a strainer and use as a mouth wash to help get rid of bad breath (page 219)."

"<u>Parsley</u>...Removing 'Dragon Breath'...Ever smell a dog's breath or someone with acute halitosis? They're bad enough to gag you. But now there's a simple cure for both extremes. The next time you feed your dog, mix several sprigs of parsley in with a little raw chopped or ground beef, then combine that with the animal's regular dry chow. You'll be surprised how well this works! And as for human breath problems, simply dip a couple of sprigs in vinegar and thoroughly chew them slowly before swallowing. The purifying effect should remove offensive odors for at least 3-4 hours (page 260)." WOW! This is said to even be used in "removing 'Dragon Breath, " even when it is, "... bad enough to gag you." This is said even to work on a dog's bad breath (remember to check with you Veterinarian for your dog before using in case of allergies, etc. of your dog)! I have tried this one with such great results that the affirmation of, "My breath need never smell bad anymore at all as long as

I give myself unconditional love, " is a reality. And beyond this recovery

is the ecstasy of knowing that a good number of people like the smell of

parsley. I came to see that all I needed to do for my bad breath was to

heal the problem the best I can; and follow that with the enjoyable

curiosity of working to honor the person I talk to with the odor that would

be delightful and ecstatic to him/her.

What a relief to the feeling of despair or hopelessness of never recovering

from toxic shaming by others because of the bad breath. Now I had the

tools to remove the bad breath and make it smell good; and to use

whatever good higher power was necessary to heal the toxic shame tapes

that in reality were not me, not part of my Iamness. I found that the Most

High Power, God, as my Highest Power, was and is the most effective and

efficient and quickest way to do this re-covery work. He gives me the

highest power or ability to conquer with might the troublesome and

untrue, and sometimes weird and haunting and compulsive and nagging,

thoughts of the toxic-shaming tapes.

"Rosemary...Effective Mouth Wash...Rosemary tea makes a

wonderfully refreshing mouth wash for getting rid of bad breath. In 1

pint of boiling water removed from the heat, steep 3 tsp. of the dried

flowering tops or leaves for half an hour, covered. Strain and refrigerate.

Gargle and rinse mouth each morning or several times a day with some (page 301)."

"*Thyme*...the French herbalist Maurice Messegue...devised various preparations for using this...for...mouth wash...bad breath...make a tea by steeping a dozen sprigs of fresh thyme in 1-3/4 pints of boiling water, covered and away from the heat for half an hour. Strain and drink 3-4 cups daily (page 338)."

"*Fennel Seed*...Breath Sweetener...If you suffer from halitosis, just chew some fennel seeds for a while and you'll have breath fresh enough for someone to kiss who may love you a lot! (page 136)" The freshness of breath is what I always wanted and now can have!

The Art of Kissing, by William Cane, First Edition, 1991, using St. Martin's Press: New York, shows the ecstasy of using a breath enhancer for kissing. "Do you like the taste of your own mouth? If not, work to make sure that your breath is sweet and fresh and that your teeth are sparkling white. Flavor your mouth with mints so that your kisses will taste sweet. Do you like the taste of your partner's mouth? If not, suggest that he or she do something about it (page 52)." It is important to note here the close relationship between what is used inside the mouth of the kisser to the resultant odor that comes from what is in the mouth. These *two go together so much to produce the ecstasy that can enhance a marriage. And when the mate gives unconditional love, the other may hear the words of affirmation that can help to heal and build him/her: "Wow! You smell great! I want to be with you! I like you!" Other affirmations can be used to help your spouse to overcome old negative tapes; thus, you can be used as a power for good! Some further quotes in William Cane's book shows how people responded in testimonials to smells in kissing: "'Apple-flavored mouths make me dizzy. There's something sensual...about the sound of your lover biting noisily into a fresh apple. The fragrance is stimulating and the taste of her mouth is so clean and bitter-sweet after she's eaten an apple that kissing her is like an aphrodisiac' (page 55)."*

And, "'I hate whiskey breath, and I refuse to kiss him when he smells like a barroom' (page 54)." More about ecstacy and kissing will be discussed when we get to the section on Skin Odors: in Recovery and in Ecstasy! There we will see what the odors of colognes and perfumes can do for people to achieve an enhanced state of ecstasy.

HEALTH IN ARMPITS:

ARMPIT RECOVERY!

Bad underarm odors can cause misery to the person possessing such a problem as well as to the people nearby. The good news is that this problem can also be conquered in most cases!

I remember when I had no odor that I could detect in my arm pits. But that was before puberty. Before puberty I would wonder why all those grown up people had to take so many bathes and/or showers. I remember thinking to myself as a child, "What a waste of time!" But I also remember thinking that there might have been something wrong with me, because I did not bath or shower as much as them.

Then about 1959 I entered puberty! A strange thing happened to me; I began getting much bigger in size, grew plenty of hair just about everywhere, and I began to notice an odor coming out from my arm pit area. Friends and non-friends alike would make remarks one to another about each other's "B.O.".

"B.O." was short for <u>b</u>ody <u>o</u>dor. They would say things like, "Put <u>it</u> down, " whenever any of us would lift our arms up into the air such as in answering a question in response to the teacher's request. This went on for years. After a while I began to think that there was something wrong with <u>me</u>; after all, wasn't I producing the odor that was making them make fun at me?

Well, the good news was that only my body was producing this odor, against my will or desire. What I am is a chooser inside my own body; an inner man that is growing through nurturing by making the right choices. I am in the image of God and am becoming that which He is, if I choose to. He told Moses to tell the people that His Name is, "I am, that I am." God calls Himself by many names such as LOVE. So being in God's image, I must do everything I can to focus my energies on becoming love. My desire is to someday say, "I am love." In 1982 a very kind man led me see what love is, and with those defining qualities I have worked diligently to become love. I do not think it fair at all to be called by anyone who may think the odor from under my arms is me, "You smell!" That is making my dignity or value of being made in the Highest Power's image, less than, or less in value than what the Creator made me worth! All of us are tremendously valuable to our Maker! That is one of the greatest truths that gives me an affirmation as to my self-worth or self-esteem that I know! Then I need to receive His Power and guidance for the "how to" become love and all the other qualities of God. We are in His image; so we must grow and be nurtured in all of His qualities.

Thus the body's odors when bad need not shame me into thinking that I am bad just like my skin odor. Rather I need to be just like, or in the image of, God doing the image of what He does; in this case, keeping the creation clean. Having a body helps me practice the qualities of God to become who I was meant to be. Throughout the creation, the unclean is cleansed when it is done by the unaltered system that was created by the Highest Power. When I choose to do what fits into the Creator's plan for me, good things happen including my developing into what I was meant to become. I like that! I have the freedom to choose my own destiny! I have decided to choose life where I can interact with everybody and everything created in the way that brings ecstasy; in fact, I hope to see a world someday that needs never recover again because recovery has been completed, where only ecstasy abounds! That is why I will keep working at my recoveries overcoming those toxic shames and other toxic qualities until victory is won and only ecstasy remains for everyone! Overcoming the odor under the arm is quite easy when a simple trick is known. The best thing I ever learned in getting rid of the smell is by use of the skin brush. The vegetable bristle variety seems to work the best for me and the friends I have told this to. The Health Food stores can be called in your area until you have found one. The method used by myself (again check with your Medical Doctor to see if you have any reason why not to use this such as very sensitive skin or lesions) is to first soak the skin brush in water for about 1/2 hour on the first day only; although, others may need to soak theirs on other days too. This will soften the vegetable bristles so that they can be used against the skin. Next, I put some good soap or shampoo on the brush's vegetable bristles, and then I start to brush or "scrub" gently under the arm.

This will remove a lot of bad smelling bacteria and dead smelly cells that the bacteria have broken down. It will also help to clean the skin probably into the pores, too. It really works for most people. Some people have told me that they do not even have to use any deodorant if they skin brush once or twice a day. I only need a deodorant if I forget to skin brush or leave my skin brush a home when I go on vacation. Later on in this book, in the chapter on Skin, I will discuss how to use the skin brush on the whole body which is wonderful!

John Heinerman in his book, Heinerman's Encyclopedia of Fruits, Vegetables, and Herbs, page 295, states, "Radishes...Deodorant for...Underarms...As a nutritious vegetable and wonderful medicinal, radishes also offer a third usefulness in the form of a toiletry for offensive body odor. But for this a juicer is needed. The juice from about 2 dozen radishes may either be put into an empty lotion bottle with a squirt top or else into a bottle with a hand-spray on it. About 1/4 tsp. of glycerine should also be added to the juice before bottling it to preserve it longer, unless you intend to refrigerate the same, in which case no glycerine is required. After your morning shower or bath, pour some of this radish juice into the palm of your hand and rub under each armpit. Or else just spray some beneath each arm...rubbing it in good to afford several hours' protection against odor."

John Heinerman gives another method of recovery from unwanted odor under the arm on page 10 of his book, "Other Uses for Cider Vinegar...Eliminates body odor when used in place of an underarm deodorant."

This leads me to one of the greatest helps I have ever learned about in all of my years of studying in the Health Field. I was told that early in our lives, my brothers and I, were being nurtured along the way by our father to the profession that he hoped would fit us best. My father, I was told, wanted to be a Medical Doctor, but he got married and never fulfilled his dream. He hoped that one of us may show a desire for the Medical field of study. He started to, along with my mother, at about the age of 3 years for myself, to teach me to be observant about healthy and sick people. I liked that! I remember studying about peas and carrots when I lived in New York at about the age of 3: I would wonder why the green color and the orange and what these different foods would do in the body.

I would experiment with the pea pods, without my parents knowing, to see if they would bring any changes in my health verses the peas themselves. I would open up the peas and see the little germ inside from which a plant could grow. I had a great time curiously experimenting with peas, carrots, and every kind of food I could get in my possession. The foundation that my father and mother gave me as nurturing parents, in a profession that I wanted to make my own, was part of the healthy functionality of my family: all families have a mixture of healthy function with unhealthy or toxic dysfunction.

To recognize the healthy parts and keep them is important on the journey to wholeness. For example, in nurturing my curiosity, creativity, and excitement in the field of health, my father and mother gave me positive affirmations that have helped me to this day. They told me that I would make a good Doctor because I am a nice person and I care about people and their health. For over 40 years now I have kept that affirmation in my heart and continue to work hard at nurturing and becoming that which I want to own as me. I have carried this healthy functionality into my present family; letting go and allowing my children, Chrissy age 14 years and Charissa age 9 1/2 years in June of 1991, to choose what they want to be in society as well as in becoming who they want to become. I work to nurture them along. I feel good with that

approach in childrearing which makes me feel that they respect the real

me, who I am, for doing so as a father. The dysfunction that I carried

into my family I am working to eliminate it: I am becoming an eliminator

of dysfunction, just like other readers who are working to become whole

in health, healthy function. And what I am that is healthy must remain

with me forever in the service of the Highest Power and His created

beings. We need all become healthy functional beings in the service of all

and ourselves in love. In order to do that we need to eliminate the

unhealthy dysfunctions out of our lives.

So what John Heinerman has mentioned in stating that cider vinegar

eliminates body odor in the pit of the arm, that physical dysfunction would

be good to eliminate. I have found (check with your Medical Doctor

before doing this incase you might have any allergies or other health

problems) that a good bath with apple cider vinegar works very well on

underarm problems, especially when coupled with the use of the skin

brush mentioned earlier under the arm. I have used 2 cups of apple cider

vinegar in a tub full of luke warm, warm, or luke warm cool water for

yearn, nearly every day. Apple cider vinegar can be purchased in the

regular grocery market. I like to buy it by the gallon as it costs less that *way and I use it so much. Apple cider vinegar has a lot of nourishment for the body as the body is being soaked in that apple cider-water mixture. It also has natural salt in it that seems to pull the physical toxins out of the body, which then pulls the smelly toxins from under the arms. The ocean water's salt seems to do this also; I had noticed during my teenage years at the beach that those that would surf or just swim a lot in the ocean did not have a smell under their arms nor did they have as many stuffed noses since the salt water probably pulled enough physical toxins out of their bodies. More will be said about this subject when we get to Skin Recovery.*

Another interesting deodorant that John Heinerman talks about is on page 304 of his book: "...Turnip...Deodorant...A Japanese delegate told me that turnip juice was one of the 'best things to use for getting rid of goaty armpits,' as he so bluntly put it. He showed me a small bottle of turnip juice which he carried everywhere with him. Being curious to see if it actually worked, I asked for and received some to use. First I washed both armpits good, then briskly rubbed 1 teaspoon of the juice beneath each of them. And although the temperature was in the high 90s and the humidity factor about 65%, yet there was virtually no odor to the perspiration. I recommend this over commercial deodorants, which contain harmful amounts of aluminum that may cause skin cancer in time. And unlike them, turnip juice won't prevent the sweat glands from doing their normal tasks, but will keep body odor from occurring for up to 10 hours as a rule."

And before leaving the discussion of underarm odor recovery, it is very important to show the importance of a good attitude. Alain Corbin (remember all these authors and their books can be found in the Bibliography at the end of this book) writes in his book on pages 38-38, "...the relation of smell to temperament...the odor peculiar to irascible personalities...passions...affected individual odor...Terror gave underarm sweat a foul smell...(also)...intolerable wind and stools." A healthy attitude appears to bring a person to better health as I explain in my book, Back To Health, that I wrote in 1979 as a First Edition. My book shows that a healthy attitude appears to relax the muscles throughout the body; and it also, therefore, takes tension off of the spinal vertebrae in such as way as to restore an improved nerve or government supply to the armpit area as well as to all the organs and structures of the body. And when nerve or government guidance of the cells, including those of the armpit, is working more completely and cooperatively, then the health of the tissues are maintained more optimally by better removal of wastes and by *better strength of the cells to fight off unpleasant odor producing bacteria.*

I thought more deeply about the elimination of offensive odors under the arm pit area when I read this account in Robert Burton's book, The Language of Smell, on page 109: "Among a certain tribe in New Guinea it is customary for friends to exchange odours when parting. A hand is pushed under the armpit, smelled and rubbed over the body. The gesture may be just a ritual but a ritual is the stilted performance of a once meaningful action...(it)...shows amicable intent by a gesture but the body odour may confirm the friendly emotion."

ECSTASY OF UNDERARM

ODORS

It feels exciting to know that even the arm pits can give off a delightful odor; that the people that we get near will actually enjoy what can come out from under the arms. I have seen a curious smile with an expression of wonderment on their faces when they get near me. I experiment with different types of neat smelling underarm applications for various reasons. For example, I may want to please myself with some kind of delightful smell that is uplifting to my thinking and feelings. Or I may want to put on the ecstatic odor of choice that my wife likes as a personal preference. Or I may want to wear the odor of choice that one of my daughters has bought for me in order to honor them as a being of creative choice. Or I may want to wear a __natural__ cologne that my "little child" remembers or likes, not only on my face but under my arms for longer lasting retention of the odor. Or I may just want to experiment with a new pleasant odor, of the many available, just to see the expressions on people's faces when I find someone who likes it. The underarms can be a great source for carrying likeable odors for a longer period of time. This is one of my favorite studies into the area beyond recovery or into ecstasy!

Now let's look into this further, if you are willing to make armpit ecstasy a part of your journey to wholeness at this time!!!

And wholeness, I feel, rightfully includes ecstasy, not just recovery! Ecstasy can become a reality for every individual as well as all of mankind as a whole. Can you imagine a world where everyone has recovered from all <u>toxic</u> problems of every kind, and only healthy problems exist such as how to even make the world and all its individuals more wholesome? It's the concept of good, better, best, and excellent. If we put aside all toxic problems and are only interested in the welfare of ourselves and all others, then we can work on improving everyone's wholeness in life. For example, if I now smell <u>better</u> by using a pleasant smelling cologne, I can improve my wholeness by finding a cologne that may mix well with my body chemistry. This would raise me from <u>better</u> to a <u>best</u> state for me. And if I get enough love and concern from my neighbors, even some professional care, I may find the <u>excellent</u> cologne for me; for example, a cologne that not only mixes with my personal chemistry but also excites me or uplifts me to feel good about myself. A <u>most excellent</u> cologne would be one that would influence me and others to serve one another in the highest most profound way possible promoting total unity, fulfilling why we were born. That would be an excellent ecstasy!!! Obviously, it will take many more things and principles to do then just put on a cologne; and I know that ecstasy can be achieved by adding and maintaining everything wholesome and right one upon another one, as many as are possible for a unique situation.

A method that I have used that works well for me is to experiment with the different colognes in the major Department Stores. They leave these out for samples for all people to test. I find that I only like certain ones. That is very important to me, that I please myself also besides my neighbor. Once I find what pleases me, then I bring my wife into the picture. I wear the colognes around her at the store. If she makes that face of ecstasy I know I have a winner. All kinds of thoughts go through my mind when I see her face light up; not just because she delights over the ecstatic odor of the cologne, but because a deeper bond is made between us that I am a person that considers her feelings and thoughts and wants besides my own. The pleasant odor is an external, and the consideration is an internal bond that can build a relationship that can last. I am looking for new ways to build greater bonds between us; and maintain the bonds that I have already built. And one more thing when discussing bonds that I feel should be mentioned: If a relationship is only built on externals like the attracting of a person by a cologne without any lasting bonding principles involved, then a person probably will only form a temporary bond that has no lasting value. In that case, the people involved usually feel eventually that they are held in bondage instead of forming a truly lasting relationship bond!

I have found that there are other relationships too that need to be built and can be by the right or fitting odor. For example, it is ok to find out what odors the people I work with like or do not like; so that I am also thinking of their welfare. What if I chose a cologne that smelled horrible to my boss or others around me? How many lasting bonds of true friendship would I make? I feel that they would feel like they were stuck at work with me in bondage; and that it would not be a positive influence for building relationships of unity for the good of everyone, rather it would be a negative influence or it would set up an environment of resistance to communicate. I need to always keep in my mind my overall big picture goal of seeing everyone united in the kind of love that will build everyone's wholeness for the benefit of everyone within the plan of the Highest Power. That is the height of excellency in my mind! Please write to me if you know of anything greater than that!

Now the cologne can be placed on the face and wrists in the right places as will be explained in the chapter on the Skin in Ecstasy. Here I want to give the technique I have found to be most helpful to me in dealing with the use of pleasant odors in the arm pit area. After skin brushing to remove all the old smelly bacteria decayed cells of the underarm area, I apply the cologne or pleasant smelling substance directly into the arm pit. I seek the most natural colognes as possible and make sure they will not stain my clothes. This area of the body seems to hold the pleasant odor longer than that of the face or wrists. (Note: I make sure I am not allergic to it. A person can check with their Medical Doctor to see if they have any allergies or other reasons why not to do this before application). This gives me a pleasant feeling inside. It makes me feel good. I like it. I also feel good knowing that I have had many previous compliments on the use of the cologne I am using. It makes me value myself more knowing that I am learning to know myself better. This creativity in experimenting to find what best <u>uniquely</u> enhances me cultivates a curiosity in me to find even more ways to heighten and magnify my being to its utmost. This I apply in every area of my life.

Now there is one more thing that need to be written about. There is a desire by some people to like the smell of their mate as a part of armpit ecstasy! This reminds me of a patient that told me that her father told her that if she did not like the smell of the man she was dating, "Don't marry *him!" Another lady told me in Big Sandy, Texas, that she just loves the smell of her husband; in fact, so much that she smells his clothes when he is not home. She said that his smell is ecstatic. (Steve Van Toller and George H. Dodd in the book called, Perfumery The Psychology and biology of Fragrance, on page 75 show that there is, "evidence for odorous steroids playing a role in axillary odour...it is conceivable that they may be somehow involved." And on page 64, "...the most used adjectives were 'urinous' and 'musky', " for the steroid found in the armpit area of men as commented on by women.) He was there in their home when she told me this in June of 1991, and he seemed to like that! I asked her to give me a testimonial about that for my book in which she told me that that would be embarrassing since talking about odors is a "Taboo". I have found this true at times. I have asked some friends why they did not give me their testimonials. They too told me they felt it was prohibited to talk about; they just felt "uncomfortable" in writing about it, to have their name in my book associated with, "such a subject." Thankfully, there are others in our culture that "would like to" write about odors they know about in their personal lives.*

And of course I need to mention that it is not the waste product odor that these people like. There seems to be something else present that identifies the person also by his/her natural odor. In fact, Robert Burton (please see the Bibliography if you want to read more on what he or others as references have written) on pages 114-115 writes about deodorants that remove or mask natural odors that can be a personal preference as being pleasurable: "The absence of a single stimulus does not greatly affect behavior and that this is the case in the human sense of smell is shown by the increasing use of deodorants and perfumes. They are designed for the removal and masking of precisely those body odours that are likely to be pheromones (pheromone comes from the two Greek words "to carry" and "to excite" and is used in the chemical sense as Robert Burton shows on page 9 of his book). As Alson Prince wrote in the Observer Magazine, the use of these products is, 'simply a form of cutting off the nose to spite the face'. It would be too much to say that the use of deodorants disturbs our behavior by removing a form a communication, as it is obvious that many users find no loss to their social life. Human pheromones may not be important but it would be a pity to abolish them. Plain food may be edible, but sauces make eating a pleasure and the same holds for other basic human activities."

Robert Burton in his book, The Language of Smell, on page 113 gives a similar account of the natural smell being liked by some: "The main *sites of sweat secretion and body odour are the armpits and the groin, which also bear the main body hair. In Man's upright posture, these areas get hot and sweaty and the hair increases the surface area from which sweat can evaporate and so promote cooling. As such the body hair could be acting as a pheromone disseminator (Greek pherein=to carry; horman=excite, page 9 of same book)...The armpits are best placed as disseminators of pheromones because they are on almost the same level as the nose. Confirmation that the armpits are a source of pheromones lies in the gesture of the New Guinea tribesmen at parting and the oft-quoted anecdote from Psychopathia sexualis by Kraft-Ebing. A young man was reputed to have great success with girls. After a dance, he would wipe the perspiring brow of his partner with a handkerchief that had been carried in his armpit. Apparently his body odour acted as an aphrodisiac and the young man claimed that his technique was highly successful. There appears to have been no attempt to substantiate this interesting technique although it opposes the usual statement that there is no such thing as a real aphrodisiac and although it is the only evidence for a male-sex pheromone in Man."*

GENITAL ODOR RECOVERY!

Unpleasant odors in the genital regions can be eliminated in most cases. Of course, a professional Medical Doctor needs to be consulted in case of any disease processes, allergies, or other types of problems that may be present. And before I begin to explain what I have learned about this area of the body, it should be noted that the groin region in which are found the genitals is one of the two sites of the body that may intensify either pleasant or unpleasant odors due to its ability to sweat. This is explained by Robert Burton in his book, The Language of Smell, on page 113, in which he states, "The main sites of sweat secretion and body odour are the armpits and the groin, which also bear the main body hair. In Man's upright posture, theses areas get hot and sweaty and the hair increases the surface area from which sweat can evaporate and so promote cooling. As such, the body hair could be acting as a pheromone disseminator." On page 9, Robert Burton gives us the definition of a pheromone: "The standard definition of a pheromone is that it is a substance which is secreted to the outside of the body and received by a second individual of the same species, in which it releases a specific reaction, for example, a definite behavior or a developmental

process...pheromones (from the Greek pherein=carry and

horman=excite." So whatever odors, pleasant or unpleasant, are found in

the armpit or groin regions, magnification of that smell will be noticed

due to the sweat __and__ the hair.

I can still hear the awful cry of my friend's girlfriend in my teenage

years saying to him, "Pooh! You stink down here!" They were in the

other room, and I thought that they were only making a phone call. I may

be wrong, but I think this was about the genitals. He was young and

probably was never taught about cleanliness in this area of his body. And

just like the arm pits, the genitals are usually not exposed to enough air

and build up more bacteria; they also seem to give off an unpleasant odor

in a number of people just like the underarms do. But the good news is

that this area can, just like the arm pit region of the body, be cleansed to

bring about a physical recovery. This in turn can help the individual that

feels ashamed of the unpleasant odor but has not known what to do about

it until the reading of this book; and it can help in mental recovery in

knowing that the problem is not about them on a mental level but rather it

was only physical in nature.

For example, if person is shamed to feel like he/she is less-than in value compared to another human being; then now they can see that they are still equally loved unconditionally by the Highest Power. There is no respect of persons to God when it comes to His unconditional love; no one is less-than another in the right and actual receiving of His unconditional love. I find I need to always keep in remembrance that I am precious and have value to God just like everyone ever created which includes you who are reading this book. I also know that I need to be nurtured and work at removing any unclean physical and mental and spiritual problems that I have. This the Highest Power delights in and gives me all the time I need to achieve it as long as I seek to overcome the pulls and pushes of my flesh and anyone who attempts to take me off of His pathways. I find that we all are never wasting our time in doing that! I also have found that to the pure all things are pure; that all things will work together in God's plan for mankind no matter what any of us do. He is in control! If we do wrong we learn; and if we do right, we learn. But so much the better if we learn doing right! The lessons of mankind are written either way! I personally do not like to toxically suffer; I realize that healthy suffering that leads me to a right way of life is important for my growth. I learn to say goodbye to toxically hurtful things like allowing my body to become unclean and not taking action to cleanse it before it harms me. I also realize that there is a proper suffering in my shortcomings and the shortcomings in others of which unconditional love works wonders! I find that by healthy modeling, we can all help each other in our journey to wholeness so that in some future time, only ecstasy will remain without the need for recovery in anyone. Giving others the unconditional love to achieve that builds love as character in us; it gets easier the more we practice and become the love we model.

One of the best ways I have ever learned in keeping the genitals clean so that they do not give off an offensive odor is to use the vegetable fiber pads and brushes as long as they are not irritating. As written in another chapter in this book, the best way to soften them is by soaking them in water for a while until they become soft enough to use comfortably. They just put some soap or shampoo on the pad and/or brush and gently clean the area. I personally like shampoo as it does not seem to leave any of the "dry" feeling on the skin that soap does. I like my body and want to treat it with respect. I have found that I first must learn to respect myself before I have the knowledge and understanding on how to treat others with respect. I cannot give something that I do not possess to give to others. Ask your Medical Doctor how to use the pad and skin brush for your particular body.

Now for the women that have to cleanse inside of the vaginal area in order to avoid having an offensive odor, a healthy vaginal douche may be what the Medical Doctor would recommends. Ask how and if you can do so in case of any allergies or other medically related reasons. Some vaginal douches can be bought at the Health Food Store. I have asked some ladies what other kinds of douches have helped them. One told me that she used diluted chlorophyll in water, while another told me she used diluted apple cider vinegar; but whatever you use, make sure your body can handle it by checking with your health professional first.

John Heinerman in his book, Heinerman's Encyclopedia of Fruits, Vegetables and Herbs, , " on page 113, states, "Coriander...Eliminates Genital Odors...In the southeastern mainland China city of Canton, coriander leaves and seeds are used to help remove unpleasant odors occurring in the genital areas of men and women...Bring 2 quarts of water to a rolling boil. Reduce heat and add 3-1/2 tbsp. of seed. Simmer for 1-1/2 hours until the amount has been reduced to slightly less than a quart of liquid. At this point, add 2 tsp. fresh, finely grated orange peel and one pitted, finely chopped date. Simmer for an additional 15 minutes, at which time remove from heat entirely. Add 1 tsp. each of dried coriander leaf (if available) and finely chopped fresh parsley, with a drop or two of peppermint oil or wintergreen oil (if available, but not necessary). Steep mixture for about half an hour, stirring occasionally. Strain through a fine sieve or filter paper and store in a pint fruit jar with a good lid to seal it. Store in refrigerator until needed. When using for genital problems, warm up whatever is needed and rub all around genital area. Let the air dry it...".

HEALTH IN THE GENITALS:

GENITAL ODOR ECSTASY

In wholesome relationships such as marriage, this can be a great blessing! Curiosity as to what ecstatic odor is being used is important. The rewards of a happier marriage can be obtained when mates know what special odors to use and which odors not to use. This is personal preference and can be used to build intimacy through showing the mate that their preferences are important. This is a part of love. Love in other languages can have more than the meaning of love in the English language. Love is not sex of and by itself; rather love can have at least three different meanings.

For example, in the Greek language, one type of love is philio or relationship type of love. Here the word is used for all of the special relationships we can have such as in the word, Philadelphia or brotherly love (philio=love; and delphia=brotherly). This is the special and unique love we need to have for our brothers and sisters. This can be how we have this philio love for our members of our family, too. How we treat our father and mother. It also can be a special and unique love between best friends like David and Jonathan in the Bible. It is a love of relationships, what kind of relationships we have, no matter what kind of relationship we have; but it is not the love that is given to everyone, it is the relationship itself, the kind of relationship. Marriage is another example of a special and unique type of love, a love that is of the philio variety. All of these philio types of love must be nurtured after they have begun their growth.

And that is where the next type of love, agape love, is so very important in order to hold together <u>any</u> relationship! It is the glue that many people have looked for but have never found! It is the bond that is the attraction that holds the people in the relationship. If that bond is made up of this agape love, it will not be broken; rather, the people in the relationship can only break up their relationship if they decide to no longer put into the relationship any more agape love. Agape love is not sex either; rather it is the fulfilling of the way of life of the Highest Power. For this, a person needs to love their neighbor as themselves. For this there are 14 qualities I learned, their is a need to be longsuffering, kind, envy not, vaunt not oneself, not puffed up, not behaving unseemly, seeking not their own, not easily provoked, thinking no evil, rejoicing not in law outside of the Highest Power but rejoicing in the Truth, bearing all Truth, believing all Truth, hoping all Truth, and enduring all Truth. This agape love is very important in my life because I find it is long lasting unlike other methods that I have used in the past that have holding power.

I go over these 14 qualities of agape love listed every day as affirmations; for example, "Ron is longsuffering and kind at the same time to others and himself!" I have learned that in order to hold onto my precious relationships I must use my Highest Power to give me the strength to be kind to those that I suffer long with. For example, if someone is not nice to me or aggravates me, I will realize that it is probably about me and my history or theirs. This helps me to see the problem in perspective and respond back in kindness. Note: there is a time to be angry in order to keep proper integrity boundaries of my identity; but I need never to forget that the person that receives my anger has their dignity or value to God. Therefore, I must do the anger in agape love. Hopefully, the person is not offended. If they are offended, I either did not use agape love to its fullest or properly, or else they offended themselves because of their own history or selves. If it was my fault, I need to ask for forgiveness.

The third type of love is eros, or aesthetic love. Here the person admires the beauty of the other person or the surroundings. This is the one type of love that many have called romantic. The candle light dinner, the beautiful flowers, the gorgeous hair, the wonderful sunset, the big muscles of the male and the curves of the female, the size of the person in every way, the beautiful way the person talks, the way one looks (he or she is so good looking), or any other pleasant characteristics of the person or place or things in the place. This has deceived more people into thinking that they have real love. When this eros love is all that they have in their love, it usually ends in disappointment due to a lack of bonding or glue to hold the two people together. It also lacks the philio love sometimes that will help hold the people together since philio love requires commitment to a longer lasting relationship!

So that is why just sex or just having an aesthetic odor is not enough. Both of these two are in the area of eros love and need to have the other two types of love, philio and agape, to make them a _love of wholeness_! Now that's _real_ love that gives _real_ true ecstasy! I have found that different levels of ecstasies can be achieved using different odors.

In his book, The Foul and the Fragrant, Alain Corbin, on page 75 in the section on, "The new Calculus of Olfactory Pleasure, " he writes, "Courtly literature was quick to record...rose water ...it was ceaselessly refreshing Conquette-Ingenue's...private parts." And on page 180 of his book, Alain Corbin writes on the importance of that "fresh" smell in his section on, *"The Perfumes of Intimacy, "* by saying, *"Fresh bodily odor depended even more on the quality and cleanliness of underwear than on scrupulous hygienic practices. Development in this area moved at an accelerated pace. Sanitary reformers endeavored to institute weekly changing of underwear."* That is a must for genital odor ecstasy! Freshness is much enjoyed by the many!

HEALTH IN THE FEET:

FEET RECOVERY FROM ODORS

How many of us have been shamed by others because our feet smelled unpleasant? I can remember when my brother would tell me that my feet "stank"! This was followed by, "pooh, go away!". I felt like I was not nearly as valuable in life as compared to him when he would make that remark so frequently. Unfortunately, I did not know how to make that awful smell go away at that time. Fortunately, I have since learned how to recover from physically unpleasant smelling feet. Further, I have learned how to recover from the shame (feeling less than what I am really worth in life): I simply put affirmations into my mind that showed that my brother was wrong, not about the smell, but about my being unworthy to stay in the area in which we were playing as children in the 1950s. I told myself that I am not required to leave just because my feet smelled unpleasant, but rather I deserve to be here as much as he. I was sometimes called "stubborn" with an "o" between the last two letters or "stubboron" as a child; but I later found out that that shaming was not always stubbornness but rather persevering in order to stand up for the maintenance of my dignity. I found that I must work at that even throughout my life if I am to have an identity of myself. This I further found needed a holding strong onto all of the qualities that are me as well as the new ones: That is my integrity or unity of qualities that make up me or what I am.

I later found through the years that followed the 1950s that simple baths in which I would wash my feet with a soap or shampoo that worked for me was a help in alleviating the unpleasant odor of my feet.

I also found that if I would change my socks more often, like once per day, I would have less unpleasant smelling feet. That may sound too simple to some people but I and a number of other people suffer or have suffered because of that practice.

I also found that if I would soak my feet when they smelled worse in water with a little bit of chlorophyll (liquid) added (about 1 tablespoon) that the smell would hardly be noticeable after a few days. This unpleasant odor would reoccur in which case I would go back to soaking my feet.

I finally found the answer for me, which may or may not work for you. (Check with your Doctor before doing this.) I used some good intestinal cleaners that I found in the Health Food store, like psyllium seed. This helped to clean out my intestinal tract. With a clean intestinal tract, the chemicals that gave the unpleasant odor were no longer great enough to travel in the blood stream to make my feet smell unpleasant. The unpleasant odor is mostly gone even to this day, June 19, 1991. I rejoice in that! I no longer worry about this problem, because I work at the maintaining a clean intestines, as far as I know, by the food I eat.

John Heinerman, on page 295 of his book, Heinerman's Encyclopedia of Fruits, Vegetables and Herbs, " gives us a look at how to help feet that have unpleasant odor: "Radishes...Deodorant for Feet...As a nutritious vegetable and wonderful medicinal, radishes also offer a third usefulness in the form of a toiletry for offensive body odor. But for this a juicer is needed. The juice from about 2 dozen radishes may either be put into an empty lotion bottle with a squirt top or else into a bottle with a hand-spray on it. About 1/4 tsp. of glycerine should also be added to the juice before bottling it to preserve it longer, unless you intend to refrigerate the same, in which case no glycerine is required. After your morning shower or bath,...just spray some...on the soles of your feet and in between the toes, rubbing it in good to afford several hours' protection against odor...".

FEET ECSTASY

WOW! How many of us have desired for everyone in the human race to have feet ecstasy? This is one of the most exciting subjects that I know. There are so many things any of us can do to make our feet feel ecstatic and smell delightful. For over 20 years now I have been working on people's feet in my offices. I find that ecstasy in the feet brings more joy in a unique and special way to the owner of the feet and their loved ones than just about any other type of ecstasy! Fortunately, when all the ecstasies mentioned in this book, along with other not mentioned in this book but I hope to write on them in future books, are combined it brings such a crescendo of joy and happiness and delight. I need to explain what I mean!

Take one of your feet into your hands if you will (Remember to always check with your Doctor first to make sure that it is ok for you to do this; and also first check for the laws in your city to see if you can massage yourself or others.). I grab my naked foot with both hands and start to gently massage the surface which will increase the circulation of the blood to the area. This will in turn bring more nutrients to make the foot healthier, and it will remove some unpleasant foot odor chemicals from the area into the blood stream. If this is done about 5 minutes a day on both feet, some pleasant changes in odor may take place. It is worth giving it a chance! (Foot massage, along with other types of family massage will be discussed in one of my upcoming books.) Also, I know how I feel great, you should feel a nice sensation of either relaxation or feeling ready to do the work you need to accomplish most of the time when this massage is done. My wife and my daughters will massage my feet at times, which not only makes me feel nice, it also builds a healthy bond in our family. I have found that if I massage them when things appear to be going wonderful, then life becomes more ecstatic; and if I massage them, using unconditional love, when things are not going so well, then our family only gets better. There seems to be a healing or healthy bonding that takes place with giving, especially giving of unconditional love where the people are unkind for the moment. It feels ecstatic to me to know the power of unconditional love!!! So the recovery is not only physical here, but it is mental and spiritual (relationships and a way of life) also!!! If you think you have a dis-eased (dis=lack of; and eased=comfort) relationship or body, then foot massage will bring back some ease to most relationships and bodies (Remember to always work with your Doctor in dealing with diseases).

For those of you who like the odor of lemons or limes or chamomile, John Heinerman, on page 105, in his book, Heinerman's Encyclopedia of Fruits, Vegetables and Herbs, " writes, "Practically everyone suffers from sore, aching feet or foot problems of some kind at one time or another. Well, there's nothing quite like a nice lemon or lime juice-chamomile cream foot rub to help ease those pains away. The feeling to be derived from something so special as this is nothing short of pure ecstacy so far as hot, tired feet and sweaty toes are concerned. Simply squeeze a little CamoCare Soothing Cream...into the palm of one hand with a little lemon or lime juice. Mix well by rubbing around in your palm with two fingers. Then rub the same on the bottoms of your soles and work well in between the toes. The sensation created is something akin to standing in ice-cold, minty water. CamoCare is distributed by Abkit, Inc. out of New York to most local health food stores. Lemons and limes are available from any produce stand or local supermarket." To combine ecstasy of these delightful odors with the ecstatic feeling in the feet is wonderful!

SKIN RECOVERY FROM ODORS

The skin is the largest organ in the body. It includes all of the areas that have been discussed except the mouth area. But it includes the arm pit area, the genital area, and the feet. What will be presently covered in this section is the recovery and ecstasy of the skin when taken as a whole. Also, special considerations of the skin such as odors of the hair and from skin problems will also be covered. (Please always remember that your Doctor needs to asked if you can do any suggestions found in this book just in case you are found to be allergic or have any other health related problems that would prevent you from using the material.)

The skin as a whole, and specific areas of it in particular, has been a part of the human body that has caused much delight when nurtured and cared for; and it has caused much heartache and grief when it has given off physical toxins that can cause unpleasant odors and lesions such as pimples that are found in acne. This in turn has caused people to shame those with the skin problems by making them feel less-than what their true value or dignity really is.

I also have been shamed by friends and others because of skin problems such as acne and rashes on my feet due to athlete's feet. I remember the unpleasant odor of the medicines used to try to remove the acne and athlete's feet which I had in my teenage years in the 1950s and 1960s. I do not remember the acne and the rashes smelling unpleasant by themselves; but I do remember those unpleasant odors in association with their hoped for removal. And I felt sad that I could not find anything else to make my acne and pimples to "go away". I did not know what I know today, about 30 years later.

For example, I have learned of the apple cider vinegar bath. What a wonderful and marvelous discovery! By using just 2 cups of apple cider vinegar, that I purchase in the regular supermarket, in a tub of luke-warm or luke-cool water, I found that some skin problems could be helped. I found that I needed to do this every day without missing a day if possible. I did it for about 15 minutes to 1 and 1/2 hours for each bathing changing the water each time. I noticed a "scurf ring" or that ring in the tub due to dirt or toxins coming off of my skin after letting the water out of the bath tub. To my surprise, my skin started looking much healthier, too. Also, my previous body odors under my arm pits (see that section in my book where I go into more detail on apple cider vinegar) began to lessen more and more especially when I would use the skin brush with it vegetable bristles. I told some of my friends and they found that their acne would either be controlled well or else be mostly eliminated. One of my friends was happy to find that his acne was arrested and that he did not smell like an Italian oil and vinegar salad like he had feared would happen. The lady who told me about the apple cider vinegar bath in New York had said that she experimented and found that 2 cups cleanses the skin and avoids the Italian salad odor. I happen to like that Italian salad smell. (Why do people go to Italian restaurants for oil and vinegar salads where they enjoy the odor; and then dislike it on the breath and skin? Is that cultural? Or is it maybe body chemistry in which, just like perfumes and colognes "combine" with body chemistry to make a pleasant aroma, the body's chemistry "combines" with the vinegar (or even garlic) to make an unpleasant odor?)

The skin also will take on the odor of some of the kinds of foods eaten! This has made me very careful in my selection of foods for present and upcoming events. For example, I will be very careful not to eat too much garlic before I take my wife out on a date as she tells me that my _whole body_ smells of garlic which she does not like. My date with her is not successful when I eat the food, garlic, that she does not like. In his book, The Foul and the Fragrant, Alain Corbin on pages 39-40 writes about whole body skin odors and food: "Regional populations exhaled specific odors, again as a result of the kind of food they ate. 'When harvest time brings these people together in our cantons, it is easy to tell men from the Quercy and Rouergue regions by the fetid odor of garlic and onion they give out, while the odor of the Auvergnats is like soured whey.' On the whole, these odors were more pronounced in southern regions."

In fact, the skin that gives off the body odor at large, was used by the Father of Medicine, Hippocrates, to help determine dis-ease states: "The odor of bodies also became an element of medical semiology. Hippocrates had already classed it as a symptom. Invasion by disease could be diagnosed both by the loss of a healthy odor and by the appearance of a morbid one. The progression to disease and then to death went from the acidic to the alkaline condition of the putrid matter...(page 40 of Alain Corbin's book, The Foul and the Fragrant)." If the reader has any unusual smell that concerns him/her, then their Doctor needs to be consulted.

In fact, I have noticed that my own personal odors have improved markedly as I have followed good sound advice of Medical Doctors, Chiropractic Doctors, and Nutritionists. For example, my father-in-law is a Medical Doctor whom I honor and respect highly for his talents not only in the field of Medicine, but in the area of nutrition. He told me to exercise in a way to improve overall health and vitality. He also told me in a non shaming way what would be best to eat and what would be best not to eat. For me and my history, my body has a tendency to get constipated and to build a high cholesterol level in the blood stream. By following his expert advice of cutting down on the number of eggs I was eating, and the amount of butter I was using on my bread; I was able to detect a noticeable change in my skin odor all over my body.

The Chiropractic Doctor was able to help me by getting the proper nerve or government supply to my skin and organs to help get rid of unpleasant body odor. The nerve supply is so important to direct the body to remove the chemicals that are causing the unpleasant odors. It also will make the organs that remove the unwanted odors to work better at their highest potential. It also will guide the nutrition that we get from our foods to whatever organs need it for nurturing so that the organs may do their work as they were designed. The function of the nervous system, that I learned and taught in Chiropractic College, is to control and coordinate all the other structures of the body; and to relate the individual to his environment (this was learned out of Gray's Anatomy, the 28th Edition, page 4, if my memory serves me correctly. I plan to write more *on the nervous system in a future book or in a future edition of this book.*

The joys of having skin that has a pleasant scent to it is simply marvelous! After recovery from odors that are not so pleasant, then to bring the body to its fullest potential of having an aroma that is pleasing is in order. For this there are many different things that can be done which include the use of colognes and perfumes, bathing in special oils and/or extracts of plants, the use of specially scented soaps and shampoos, etc. And what makes this so exciting is that there is so much to choose from. And any scent that is available can be used to fit a purpose or just for enjoyment because an odor is personally preferred. Creativity can be nurtured here! Different combinations can also be put together! (But remember to be checked by your Medical Doctor for allergies or other health related problems before doing these.) These different combinations put together creatively are commented on by the great Father of Medicine, Hippocrates, in Alain Corbin's book, The Foul and the Fragrant, on page 13 where Mr. Corbin says, "...in the fifth and fourth centuries B.C. Hippocrates and his disciples at Kos had already emphasized the influence of air on fetal development, the formation of temperaments, the

birth of passions, the forms of language, and the spirit of nations...".
That is pretty significant: the odors we select not only influence ourselves
but others in development and formations of their being!

And of course, as quite a few people have told me privately as they see
its effect in their lives and others, and has been expressed in this letter
from an admirerer in Alain Corbin's book, *The Foul and the Fragrant*, on
page 207, "August 15: Tell me if you use verbena; do you not put it on
your handkerchiefs? Put some on your slip. But no-do not use perfume,
the best perfume is yourself, your own fragrance." Some people have told
me that the best perfume is their own mate; and further, they told me that
when they can combine the perfume of their mate with a perfume or
pleasant odor that "mixes well with their mate's body chemistry", then
that is a special WOW!!! And Steve Van Toller and George H. Dodd on
page 5 write: "Daly and White, writing in 1930, noted that the functional
significance of perfume may not be for the purposes of disguising or
masking natural body odour, as is widely thought, but to heighten and
fortify natural odour."

Even George Armerding on page 45 of his book, *The Fragrance of the
Lord*, shows how special it is to have a mate that smells pleasant: "Even
the patriarch, Abraham, could not escape the influence of fragrance.

When Abraham was an old man, and his body in a manner dead, he took to wife a woman named Katurah (Gen. 25:1). By interpretation her name means 'incense' or a 'sweet odor.' To put it bluntly, Abraham obtained a delightful fragrance through marriage. There is real satisfaction in a marriage like that. We do not often think of marriage as bringing with it a fragrance, but such a circumstance provides a real enrichment to any life." Any man who loves the scent of his wife would agree to this! And just think how it would be if her beautiful smell was creatively mixed with perfumes that blend delightfully with her scent! What heights of feelings and emotions of ecstasy could be achieved? If the husband treated her with the dignity and respect she deserves as a creation of God at the same time, the level of ecstasy may be indescribable! And if the wife did the same for the husband, then WOW!!! The more I combine the physical odors with the mental and spiritual odors, the aroma only gets better and better!

This is just like the origin of the "toast" given between people to honor someone. As they hold up their glasses of wine and "clink" glass to glass, they say, "I want to give a toast to....". The origin is supposed to be a blending or mixing of the flavors of the wine with the toasted bread with the person. The person had a flavor all of his own that gave off an aroma, such as generosity or pleasantness or helpful, that was needed or vital to the people toasting.

That same harmonizing of odors is what makes the _art_ of skin ecstasy so delightful, exciting, and provoking to creativity! I have noticed that the art of skin ecstasy is to find that special one or many colognes, oils, and herbal smells that mix very well with my body chemistry to make me smell pleasant in a special and influencing way of my choosing for any occasion I desire! And Alain Corbin, on page 204, puts it in perspective to show the importance of our awareness of our own selves even in the area of odors, "Since scientists laid it down that every individual possessed a specific odor, smelling oneself, examining the changes in the smells of one's own body, was already to become aware of the nature of one's being." And my being aware of my own being lets me know who and what I am so that I know my limits of influence on myself and others.

The smell that I give off tells of my internal environment as well as my externals. If we have an attitude that has an unpleasantness to it, then our very body will give off an odor that is unpleasant; and if we have a pleasant way about us in approaching life, then our odor will be more pleasant. I have noticed that when I have an unpleasant attitude, I then have a change in body odor toward the unpleasant; and when I am doing the attitude of becoming more and more like my Maker with genuine love for everyone, then I find that my body smells more pleasant. Alain Corbin shows this on pages 38-39 of his book: "Oddly, medical discourse concerned itself little with the specific relation of smell to temperament, the color of hair, or complexion. The odor peculiar to irascible personalities and the smell of redheads were noted, but without emphasis, as if they were self-evident. Passions exercised an effect on the humors; they also affected individual odor. Some passions operated slowly but profoundly; they checked organic movements and cut off secretions. Thus people lost their odor when they were sad. Passions that struck by fits and starts intensified the bodily stenches. Accelerated putrefaction of the bile reinforced the smell of breath when a person was angry. Terror gave underarm sweat a foul smell and created intolerable wind and stools."

Robert Burton shows, on pages 109-110, this importance of learning temperance or self-control over our emotions if we are to have healthy smelling skin: " There are also stories of the emotion of fear being communicated by smell. Some people claim that they can smell fear in others..."

Alain Corbin further shows on page 75 that, "The persistent emphasis on the skin's absorptive power justified the caution regarding the use of strong scents. This caution, however, was counterbalanced by the use of perfumed powder, which, more than other cosmetics, revealed the personality of the person using it. It 'varies,' Dejean noted, 'according to each person's taste and is composed of distinctive perfumes.'...Carnation powder came to the fore in the reign of Louis XV, its success signified the triumph of vegetable scents." It is interesting to note that even the personality of the individual is revealed by the type of scent used. I find that I like to use colognes that mix well with my body chemistry but are not seductive, only to make an atmosphere of happiness about me for me and the people near me. I try to make it a mixture that will not offend anyone; but rather, a mixture that will help to bond me to people in a way that brings healthy conversation for growth and nurturing.

And George Armerding in his book shows in his Foreword, "Part of the reason for the unique place odor has in our lives is the great variety of responses the same perfume evokes when smelled by different people. In one it may bring back the memory of a vacation in childhood, in another a hike beside a mountain stream, in still another a stroll in a garden with a lover. Thus a perfume can become a very personal and subjective experience, difficult, if not impossible, to explain. We tend to like those *odors that conjure up pleasant memories and to shun those that do not...".*

And as Steve Van Toller and George H. Dodd say on page 185: "The

correlation between personality and fragrance was initially a supposition,

but the latest research in the psycho-physiological area supports this

idea." This is important as it shows me that I am an individual; and

because of my identity being so unique, I will subjectively respond

differently than anyone else according to my characteristic life

experiences. And because of that, I am cautious but curious as to the

responses of people around me to my colognes. It is exciting to ask people

what they feel and think about the cologne I am wearing. In fact this is

on the minds of more people than I thought as shown in the book,

Perfumery The Psychology and Biology of Fragrance, on pages 205-206:

"In an American survey we asked nearly 800 women which two of a

number of personal attributes they noticed about people on first meeting.

To our surprise, 43 per cent indicated 'smell'; slightly more indicated

'face', 'eyes' and 'voice', but fewer talked of 'hair', 'dress', 'skin',and

'hands'. So it is hardly surprising that motivational research reveals that

women are very concerned with the messages they send out about

themselves when they use perfume."

Culturally, I found, as I have travelled nearly all over the world, that I need not worry about my body smelling pleasant to the people in a

particular land if I eat what they eat. That is not personal preference, but

rather it is cultural preference! If I am to smell more ecstatic where I

travel, I had better do as they do in terms of food consumption. What

smells pleasurable in one country is unpleasurable in another sometimes.

Also, in America, I lived nearly 43 years out of my present 44 years (today

is June 27, 1991) in the state of California, in the general area of the city

of Los Angeles, where people come from all different countries and now

live there. There I had noticed that they had different body odors

something like Robert Burton talks about in his book on page 113 as

follows: "Sweat is a more likely carrier of pheromones. Sweat glands are

found all over the body and sweat carries the characteristic body odour.

To some extent the body odour (excluding breath odour) is related to diet.

The odour of Eskimos is due to a diet of fish; Mediterranean people smell

of garlic and onions and northern Europeans have a cheesy, buttery

smell. Within these broad categories there is much variation and people

with a good sense of smell can identify individuals by odour alone, using

true body odour and not tobacco, perfume or other acquired odours."

BIBLIOGRAPHY

Armerding, George D., The Fragrance of the Lord, Harper and Row: San Francisco, First Edition, 1979.

Burton, Robert, The Language of Smell, Routledge and Kegan Paul Ltd: London, First Edition, 1976.

Cane, William, The Art of Kissing, St. Martin's Press: New York, First Edition, 1991.

Corbin, Alain, The Foul and the Fragrant, Harvard University Press: Cambridge, Massachusetts, First Edition, 1986.

Duskis, D.C., Dr. Ronald Alan, Back To Health The M.A.N. Principle, Herrick House: Monrovia, California, Second Edition, November of 1979.

Gray, Robert, Gray's Anatomy, 28th Edition.

Heinerman, John, Heinerman's Encyclopedia of Fruits, Vegetables and Herbs, Parker Publishing Company: West Nyack, New York, First Edition, 1988.

Morris, Edwin T., Fragrance The Story of Perfume from Cleopatra to Chanel, Charles Scribner's Sons: New York, First Edition, 1984.

Van Toller, Steven, and George H. Dodd, Perfumery The Psychology and Biology of Fragrance, Chapman and Hall Ltd.: New York, First Edition, 1988.